THE

METALLIC WEALTH

OF

THE UNITED STATES.

THE

METALLIC WEALTH

OF

THE UNITED STATES,

DESCRIBED AND COMPARED WITH THAT OF OTHER COUNTRIES.

BY

J. D. WHITNEY.

PHILADELPHIA:
LIPPINCOTT, GRAMBO & CO.
LONDON: TRÜBNER & CO.
1854.

LIST OF ILLUSTRATIONS.

WOOD-CUTS.

1. Section of a stockwerk deposit.
2. Contact deposit of ore between two formations.
3. Section of segregated veins.
4. Section of the Rammelsberg, from Burat.
5. Section illustrating the nature of gash-veins.
6. Part of the lode at Wheal Julia, to illustrate the comby structure of lodes, from De la Beche.
7. Fragment of the Drei Prinzen Spat Vein.
8. Ideal transverse section, to illustrate the terms used by miners in speaking of a vein.
9. Section of a vein at Holzappel.
10. Section of a vein, with branches coinciding with the bedding of the rock.
11. Ground plan of the intersection of veins in the Himmelfahrt Mine, near Freiberg, from Weissenbach.
12. Transverse section of a shaft and adit-level.
13. Ideal section of shafts for opening a mine.
14. Horse-whim.
15. Method of overhand stoping.
16. Timbering in the levels of a mine.
17. Timbering in mines, from Burat.
18. Timbering in levels, with solid rock on one side.
19. Section of the junction of the greenstone and amygdaloid on Keweenaw Point.
20. Section of the Hogan Mine, March 1854.
21. Section of the Northwestern Mine, February 1854.
22. Section of the South Cliff Mine, February 1854.
23. Mass of copper containing nodules of Prehnite.
24. Section of the Siskawit Mine.
25. Deposit of copper between sandstone and trap in the Porcupine Mountains.
26. Parallel layers of cupriferous veinstone in the trap, at the Douglass Houghton Mine.
27. Section of the Douglass Houghton Mine, February 1854.

28. Section of the Toltec Mine, March 1854.
29. Transverse section of adit and shaft, No. 3, Minnesota Mine.
30. Section of workings on the South Lode, Minnesota Mine.
31. Section of Norwich Mine, February 1854.
32. Section of the Dolly-Hide cupriferous bed.
33. Section of the M'Cullock Vein.
34. Section of the East Tennessee copper veins.
35. Occurrence of copper ores in the New Red Sandstone.
36. Section of the Zinc and Franklinite beds at Stirling, New Jersey.
37. Section of the Coal Hill Mine.
38. Section of the workings on the Union Vein.
39. Section of the Middletown Silver-Lead Mine.
40. Branching of the lode at the Middletown Mine.
41. Section of the Washington Mine, North Carolina.
42. Ideal section of a Lead fissure, in the western lead-region.
43. Horizontal deposit of lead ore.
44. Section of Levins's Lead near Dubuque.

LITHOGRAPHS.

Plate I. Section of the Northwest Mine.
II. Section of the Cliff Mine.
III. Section of the Minnesota Mine.

NOTE.—All the sections are drawn on the scale of 200 feet to 1 inch. The stopings are represented by the part in black.

CONTENTS.

INTRODUCTION.

Sketch of the history of mining in the United States; mining speculations; object of the present work; its arrangement.

CHAPTER I.

ON THE NATURE OF THE DEPOSITS OF THE METALS AND THEIR ORES, AND THE GENERAL PRINCIPLES ON WHICH MINING IS CONDUCTED.

Importance of the subject, 33; classification of metalliferous deposits, 34; superficial deposits, 34; stratified deposits: intercalated, disseminated, metamorphosed, 35; theories of their origin, 36; unstratified deposits: division of, 36; irregular deposits: eruptive masses, 36; how mined, 37; dissemination through eruptive rocks, 38; washings derived from their disintegration, 38; stockwerk deposits, 39; instances of, 39; contact deposits, 40; instances of, 40; their character and value, 41; fahlbands, 42; regular deposits, 43; definition of a vein, 44; Wissenbach's classification of veins, 44; segregated veins, 45; how different from true veins, 46; instance: the Rammelsberg, 47; their character and value, 47; gash-veins, 48; their origin, 48; how filled, 49; true veins: definition, 49; how originated, 49; various dimensions, 50; contents, 51; occurrence in groups, 51; veinstones, 51; their comby structure, 52; remarkable instance, 53; various vein-phenomena, 54; selvages, 54; ideal section of a vein, 55; worthless matter contained in veins, 55; their varying richness, 56; change of direction, 56; branching, 57; derangements from intersection, 57; theories of the origin of mineral veins, 59; theory of accidental formation, 60; of igneous injection, 60; of sublimation, 61; of aqueous deposition, 62; of lateral secretion, 62; discussion of circumstances contributing to the formation of a lode, 63; electro-chemical action, 65; decomposition of upper portion of lodes, 66; variations in character at different depths, 67; general principles of working of mines, 68;

adit-level, 69; removal of water, 69; shafts, 71, whether vertical or on the lode, 71; raising the ore, 73; levels, 74; stoping, 75; timbering, 77; mining of masses, 77.

CHAPTER II.

GOLD, PLATINA, AND SILVER (IN PART).

SECTION I.

MINERALOGICAL OCCURRENCE AND GEOLOGICAL POSITION OF GOLD.

Mineralogical occurrence: forms of occurrence, 79; native gold, 79; alloyed with silver, 79; analyses of, 80; occurrence of gold compared with that of other metals, 81; alloyed with iron, 82; form, 82; geological position: confined to lowest fossiliferous strata, 82; general resemblance of gold-bearing rocks everywhere, 83; gold-veins of segregated class, 83; rarity in unaltered rocks, 84; period of origin of the veins, 85; gold actually obtained chiefly from washings, 86; its separation and concentration, 86; whether due to still-existing causes, 87.

SECTION II.

GENERAL DESCRIPTION OF FOREIGN GOLD REGIONS.

Geographical order followed in the work, 88; gold region of the Ural and Siberia, 88; Great Britain, 90; recent gold-excitement in, 91; gold-quartz machines, 91; Austria: Hungary and Transylvania, 92; Tyrol and Salzburg, 93; Bohemia, 93; proportional produce of the different provinces, 94; France: the Rhine, 94; Spain, 95; Italy, 96; Central Asia, 96; Southern Asia, 97; China and Japan, 97; East India Islands, 98; Africa, 98; Australia: general geology of gold region, 99; history of discovery of gold, 100; its quality, 102; nature of occurrence, 103; origin of the deposits, 104; statistics of produce, 104; South America, 108; New Grenada, 109; English companies working in, 110; Venezuela, 110; Brazil, 110; geology of mining region, 111; St. John del Rey Mine, 111; Mexico, 113; Central America, 114.

SECTION III.

GEOGRAPHICAL DISTRIBUTION OF GOLD IN THE UNITED STATES.

Two distinct gold regions, 114; stimulating effect of California discoveries, 114; history of discovery of gold and earliest workings in Southern States,

115; North Carolina, 115; Georgia, 118; South Carolina, 118; Virginia, gold mines in 1836, 119; South Carolina in 1848, 120; description of Appalachian gold district, 121; gold in Canada, 123; Vermont, 123; Maryland, 124; Virginia, 125; companies now working in: Culpeper, 125; Freehold, Liberty, Gardiner, Marshall, Whitehall, 126; Waller, Garnett, and Moseley, 127; London and Virginia, Buckingham, 128; North Carolina, 129; companies now working in: Gold Hill, M'Cullock, 129; Conrad Hill, Vanderburg, 130; Phœnix, Long and Muse's, 131; Lemmond, Capp's, Mecklenburg, 132; South Carolina, 133; Dorn Mine, Dorn Mining Company, 133; Georgia, 134; Tennessee and Alabama, 134; New Mexico, 134; California, 134; history of discovery of gold, 135; general description of the gold region and its geology, 137; age of gold-bearing slates, 139; occurrence of the gold in washings, 140; in the rock, 141; gold-quartz mining companies, 142; hydraulic works, 143; machines used, 143; Mint statistics of gold produced in United States, 145; statistics of produce of California, 146; general statistics of gold production: eastern hemisphere, 147; western, 148; comparative production at different periods, 149.

SECTION IV.

PLATINA, AND ITS ASSOCIATED METALS.

Occurrence of platina, 149; history and use, 150; alloys of other metals with, 150; geological position, 151; geographical distribution: Russia, 152; France, 153; Germany, Pettenkofer's experiments, 153; Spain, 154; East India, 154; Borneo, 154; South America, 155; Canada, 156; United States, 156; platina in California gold, 156; importation into this country, 157.

CHAPTER III.

SILVER.

SECTION I.

THE MINERALOGICAL OCCURRENCE AND GEOLOGICAL POSITION OF SILVER.

Mineralogical occurrence: native metal and alloys, 158; amalgam, 159; ores, 159; geological position: general diffusion, 160; sources of silver of commerce, 160; veins and masses, 161.

SECTION II.

GEOGRAPHICAL DISTRIBUTION OF SILVER ORE.

Norway: Kongsberg Mine, 162; Saxony and Bohemia, 163; Freiberg mines, 164; Austrian Empire, 166; proportional yield of the different provinces,

167; Hungary, 167; gigantic adit, 168; South America, 169; Peru: mines of Pasco, 169; Bolivia: mines of Potosi, 171; Chili: geology and mining districts, 172; mines of Copiapo, 174; Mexico, 175; sources of our knowledge, 175; nature and position of veins, 176; English mining companies, 178; causes of their ill-success, 179; United States, 180; general statistics of produce of silver: in eastern hemisphere, 181; western, 182; general comparative statement, 183; relative production of gold and silver, 184; silver as a standard of value, 185.

CHAPTER IV.

MERCURY.

SECTION I.

MINERALOGICAL OCCURRENCE AND GEOLOGICAL POSITION OF THE ORES OF MERCURY.

Mineralogical occurrence: native metal, 186; amalgam, 186; ores, 186; geological position, 187.

SECTION II.

GEOGRAPHICAL DISTRIBUTION OF MERCURY AND ITS ORES.

France, 188; Bavaria, 188; Austria, 188; Spain: antiquity of workings, 189; nature of deposits, 189; yield, 190; Tuscany, 190; South America: Peru, 191; mine of Santa Barbara, 191; occurrence of native mercury, 192; New Grenada, 193; Mexico, 193; consumption of mercury in, 194; United States, 195; New Almaden Mine, 195; statistics of produce of mercury, 197; estimate of present production, 197; importation into this country, 197.

CHAPTER V.

TIN.

SECTION I.

MINERALOGICAL OCCURRENCE AND GEOLOGICAL POSITION OF TIN.

Mineralogical occurrence: native tin, 198; ores, 198; sources of supply of tin, 199; geological position: general mode and place of occurrence, 199; forms of deposits, 199; occurrence of certain metals in washings, 200; minerals accompanying tin, 201.

SECTION II.

GEOGRAPHICAL DISTRIBUTION OF TIN.

Great Britain, 201; general geology of Cornish mining region, 202; veins and fissures in, 204; occurrence of tin in Cornwall, 206: produce, 207; Saxony, 209; Austria, 210; France, 210; Spain, 210; Malay Archipelago: Banca, 210; Malay Peninsula, 212; South America, 212; Mexico, 212; United States, 213; general statistics of produce of tin, 213; tin commerce of Great Britain, 214; consumption of tin in United States, 214; note on present produce of Great Britain, 215.

CHAPTER VI.

COPPER.

SECTION I.

MINERALOGICAL OCCURRENCE AND GEOLOGICAL POSITION OF THE ORES OF COPPER.

Mineralogical occurrence: general diffusion and early use, 216; native metal, 216; ores: combinations with sulphur, &c., 217; with oxygen and chlorine, 218; silicates, 219; carbonates and sulphates, 220; geological position, 220.

SECTION II.

GEOGRAPHICAL DISTRIBUTION OF COPPER IN FOREIGN COUNTRIES.

Universal distribution, 222; Russian Empire, 222; Norway and Sweden, 224; Great Britain, 226; decomposition of upper part of veins: gossan, 226; ores lower down, 227; influence of nature of enclosing rock, 229; variation in character in depth, 229; great Cornish adit, 230; history of different mines: Dolcoath, Great Consolidated, Devon Great Consolidated, 231; history of copper mining in Cornwall, 232; statistics of produce, 233; total production of Great Britain, 235; Prussia, 235; Austrian Empire, 236; France, 237; Spain, 238; Italy, 238; Turkey, 238; Algiers, 239; East India, 239; Japan, 239; Australia: Burra-Burra mine, 239; Kapunda, 240; New Zealand, 241; South America: Peru, 241; Chili, 242; imports into the United States from, 243; Cuba, 243; Cobre and Santiago mines, 244; imports into the United States from, 245; Jamaica, 245; Mexico, 245.

SECTION III.

GEOGRAPHICAL DISTRIBUTION OF COPPER IN THE UNITED STATES.

Importance of copper in the United States, 245; classification of deposits of ores, 246; copper region of Lake Superior: history of discovery, 247; progress of exploration and development, 248; speculations, 249; geological survey, 249; ancient mining, 250; general geology of the region, 251; trap ranges, 252; character of the metalliferous product, 254; division of the mines into groups, 255; Keweenaw Point mining region: geological description, 255; character of the veins, 258; form of occurrence of the copper, 260; description of mines in the district: New York and Michigan, Clark, Washington, Agate Harbor, 262; Native Copper, Eagle Harbor, Copper Falls, 263; Humboldt, Meadow, Phœnix, 266; Keweenaw Point, Star, Manitou, 268; Iron City, Bluff, Northwest, 269; Cónnecticut, Waterbury, 272; Summit, Dana, Northwestern, 273; Winthrop, Eagle River, 275; Pittsburgh and Boston: discovery of Cliff Vein, 275; workings, 277; occurrence of silver, 278; North American: old mine, 279; new, or South Cliff mine, 280; Albion, Fulton, 281; Manhattan, Montreal locations, 282; Isle Royale: general description, 283; mines upon: Siskawit, 284; Pittsburgh and Isle Royale, 285; Ontonagon mining district: general description, 285; mode of occurrence of the copper, 286; forms of deposit, 287: description of mines in the district: Algonquin, 289; Douglass Houghton, Toltec, 290; Algomah, Aztec, 291; Bohemian, Ohio, Adventure, 292; Ridge, Evergreen Bluff, 293; Minnesota, 293; south lode, 295; Rockland, Flint-Steel River, Peninsula, 296; National, Forest, 297; Glen, Norwich, 298; Windsor, Ohio Trap-Rock, Sharon, 300; Portage Lake district: mode of occurrence of the copper, 301; description of mines: Isle Royale, Portage, 302; Montezuma, Albion, Quincy, 303; produce of all the Lake Superior mines, 303; amount invested in them, 305; their value, 305; mining on north shore, 305; on north shore of Lake Huron: age of copper-bearing formation, 307; Bruce mine, 307; copper deposits of the Mississippi Valley: geological position, 308; Wisconsin, 309; Missouri, 310; Mine La Motte, 310; Current River region, 311; cupriferous deposits of the Atlantic States: in metamorphic rocks, 312; their mode of occurrence, 312; Maine, 313; New Hampshire, 313; Vermont, 314; Massachusetts, 314; Connecticut: Bristol mine, 315; New York, 316; Pennsylvania, 316; Maryland: mines, New London, Dolly-Hide, 317; Springfield, Mineral Hill, 319; Virginia: Manassas mine, 320; North Carolina: North Carolina mine, 320; copper in the gold mines, 321; Tennessee: character of deposits, 322; gossan and black ore, 322; dimensions of veins, 323; companies working, 324; copper ores in New Red Sandstone, 324; mode of occurrence, 325; Connecticut: Simsbury mines, 325; New Jersey, 326; history of earlier mining operations, 326; copper fever in 1847, 328; copper at junction of gneiss and New Red Sandstone; Pennsylvania, 328; Perkiomen and Ecton mines, 329; copper mines of New Mexico, 330; general statistics of production of copper: eastern hemisphere, 331; western, 332; comparative statement, 333; importations and exportations of copper, &c., in this country, 333; smelting of copper, 334; foreign ores smelted at Swansea, 335; later history of the business, 335; smelting establishments of this country, 336.

CHAPTER VII.

ZINC

SECTION I.

MINERALOGICAL OCCURRENCE AND GEOLOGICAL POSITION OF THE ORES OF ZINC.

Mineralogical occurrence, recent introduction into use in form of metal, 337; ores: combinations with sulphur, 337; with oxygen, 338; silicates, &c., 338; geological position: general diffusion, 339; in regular veins, 339; in beds and masses, 340.

SECTION II.

GEOGRAPHICAL DISTRIBUTION OF THE ORES OF ZINC IN FOREIGN COUNTRIES.

Poland, 340; Great Britain, 340; Belgium, geological formations of, 341; occurrence of calamine in, 342; Vieille Montagne works, 345; Nouvelle Montagne, 343; produce, 344; Prussia: metalliferous district of Silesia, 344; Westphalia and Rhine, 345; Austria, 346.

SECTION III.

DISTRIBUTION OF THE ORES OF ZINC IN THE UNITED STATES.

New Hampshire, 347; New York, 347; New Jersey: position of ore-beds, 348; New Jersey Zinc Company's works, 350; Pennsylvania, 351; Pennsylvania and Lehigh Zinc Company, 351; Western lead region, 352; general statistics of production of zinc, 353; use of the oxide as paint, 353.

CHAPTER VIII.

LEAD, .ND SILVER IN PART.

SECTION I.

MINERALOGICAL OCCURRENCE AND GEOLOGICAL POSITION OF THE ORES OF LEAD.

Mineralogical occurrence: native metal, 354; combinations with sulphur, &c., 354; with oxygen, 355; with chlorine, 356; oxygen salts, 356; geological position, 357: decomposition of ores at surface, 357; association with zinc and silver, 358; source of supply hitherto, 358.

SECTION II.

DISTRIBUTION OF THE ORES OF LEAD IN FOREIGN COUNTRIES.

Russian Empire, 359 ; Sweden, 360 ; Great Britain : its importance as a lead-producing country, 360 ; North of England mining district, 361 ; Derbyshire, 362 ; Cornwall and Devon, 363 ; East Wheal Rose Mine, 364 ; Cardiganshire and Montgomeryshire, 365 ; mines of Loginlas and Goginan, 365 ; yield of British mines, 366 ; yield of silver, 367 ; Belgium, 367 ; Prussia, 368 ; Harz : its interest as a mining district, 369 ; geology, 369 ; systems of veins, 369 ; history of mining in, 371 ; produce, 371 ; Saxony, 372 ; Nassau, 372 ; Baden and Würtemberg, 373 ; Austrian Empire, 373 ; Spain, history of mining in, 375 ; character of deposits, 377 ; mines of Almagrera, 377 ; remarks on Andalusian mines, 378 ; English mining about Linares, 379 ; Italy, 380 ; France, 380 ; Australia, 381.

SECTION III.

GEOGRAPHICAL DISTRIBUTION OF THE ORES OF LEAD IN THE UNITED STATES.

General diffusion and importance of lead-deposits; 381 ; their classification, 382 ; mines of St. Lawrence County, New York, 382 ; position and character of deposits, 383 ; Coal Hill Mine, 384 ; Victoria mines, 386 ; St. Lawrence, 387 ; lead-bearing veins of the metamorphic palæozoic, 387 ; reasons why not profitably mined, 387 ; Maine, 388 ; Vermont, 388 ; New Hampshire, 388 ; mines : Eaton, Shelburne, 389 ; Massachusetts : mines of Northampton and vicinity, 390 ; Connecticut, 392 ; Middletown Mine, 393 ; New York, 394 ; Pennsylvania, 396 ; mines : Chester County, Wheatley, 397 ; Brookdale, Charlestown, 398 ; North Carolina, 398 ; Washington Mine, 399 ; lead mines in unaltered palæozoic rocks, 400 ; various deposits in New York, 401 ; Ulster County mines, 401 ; lead deposits of the Upper Mississippi, 403 ; history of their development, 404 ; extent of the lead-region, 405 ; its geology, 406 ; position and character of the lead-bearing formation, 407 ; mode of occurrence of the lead, 410 ; its quality, 412 ; various forms of deposits, 412 ; veinstone, 4[illegible] ; Levins's lead, 414 ; theory of deposition of the ore, 415 ; practical i[illegible]rences, 416 ; lead-region of Missouri, history of mining in, 417 ; its geology, 418 ; character of the deposits, 419 ; smelting, 419 ; statistics of yield of all the mines, 420 ; other lead deposits, 421 ; general statistics of produce of lead, 422 ; comparative production of different countries, 423 ; imports and exports of this country, 423 ; sources of future supply, 424.

CHAPTER IX.

IRON.

SECTION I.

MINERALOGICAL OCCURRENCE AND GEOLOGICAL POSITION OF THE ORES OF IRON.

Mineralogical occurrence: general mode of, 425; universal distribution and early use of iron, 425; native iron, 426; ores: combined with sulphur, &c., 426; with oxygen, 427; silicates, 429; carbonates, &c., 429; geological position, 429; general classification of iron ores, 430; unstratified ores: in the azoic, 430; ores of Sweden and Norway: Durocher's classification of, 431; their geological place and character, 431: eruptive deposits in this country, 433; beds apparently stratified, 434; iron-producing character of azoic age, 434; eruptive ores in more recent formations, examples of, in the Ural, 435; in Elba, 436; iron ores in the metamorphic formations, 437; stratified ores: their character and value, 438; division of the class, 438; ores in non-coalbearing rocks, 439; below the coal, 439; above it, 440; ores in the coal-measures, 442; iron-stones, 442; black-bands, 444; iron ores in tertiary and alluvial formations, 444; ores of France, 445; bog ores, 446.

SECTION II.

STATISTICS OF IRON IN FOREIGN COUNTRIES.

Russia, 446; Scandinavia, 447; Great Britain, 447; sources of information, 448; history of iron manufacture, 448; introduction of steam-engine, 449; of hot blast, 449; of use of black-bands, 450; effect of demand for railways, 451; statistics of produce and trade, 452; Belgium, 453; Prussia, 454; Austria, 454; other states of Germany, 455; France, 456; Spain, 457; Italy, 457; Switzerland, 457.

SECTION III.

DISTRIBUTION OF THE ORES OF IRON IN THIS COUNTRY.

Circumstances affecting the production of iron, 457; iron ores in Maine, 458; New Hampshire, 459; Vermont, 460; Massachusetts, 460; tertiary ores of western part of the state, 461; Connecticut, 462; New York: eastern part, 463; southeastern, 464; northern azoic region, 465; New Jersey, 467; manufacture in Morris County, 468; Pennsylvania, 469; statistics of manufacture, 470; Maryland, 472; Virginia, 473; North Carolina, 474; South Carolina, 474; Georgia, 475; Alabama, 475; Tennessee, 475; Kentucky, 476; Ohio, 476; Michigan, 477; Lake Superior iron region, 477; Indiana and Illinois, 478; Missouri, 478; Iron Mountain, 479; Pilot Knob, 480; other localities, 481; Iowa, 481; Wisconsin, 481; British provinces, 482; general summary, 483; resources of this country in coal, 486; production of iron at different periods, 487; general statistics of the production of iron, 488; comparative statement of present production, 488; our importation, 489; exportation, 490.

CHAPTER X.

METALS NOT USED IN THEIR SIMPLE METALLIC FORM.

Classification, 491; Bismuth, 491; its forms of occurrence, 491; use, 492; where found in this country, 493; antimony, 493; its forms of occurrence, 493; sources of supply, 494; use, 494; Nickel, 495; its forms of occurrence, 495; use, 496; whence supplied, 496; mine of, at Chatham, 496; Cobalt, 497; its forms of occurrence, 497; mine of, in Norway, 498; where found in this country, 499; Arsenic, 500; Manganese, 501; its forms of occurrence, 501; sources of supply, 502; use, 502; Chromium, 502; Titanium, 503; Molybdenum, 503; Uranium, 504; Tungsten, 504.

CHAPTER XI.

GENERAL SUMMARY.

Method pursued in the work, 505; tables of present production of metals throughout the world, 506; remarks on the statistical tables, 508; comparison of the total production of different countries, 510.

INTRODUCTION.

The records of mining on the American Continent form by no means the least interesting chapter in its history. The expectation of finding a land possessing greater treasures of gold and silver than existed elsewhere, led to the discovery of the New World, and was the predominating motive in its settlement. It was only to the Atlantic coast of the northern half of the Continent, that colonists came who were inspired by any higher aim than that of finding a recompense for the toils and dangers they had undergone, in the rich mines of the precious metals which they were to discover. These golden dreams were not destined to be disappointed, for a land had indeed been reached whose realities of metallic wealth were capable of satisfying an almost boundless cupidity. Under Spanish dominion, the mines of Mexico and South America poured forth their treasures of silver and gold, and deluged Europe with an amount of these metals far beyond anything ever before dreamed of. The history of Spanish mining in America is too well known to require recapitulation here. Confined almost entirely to the metals styled precious, it conferred neither happiness nor prosperity on either country.

The mines of South America had been worked for nearly a century before the first settlements took place upon the less attractive shores of the Northern Atlantic. Mining formed no part of the object of the colonists of that region, and it was only at rare intervals that some adventurous person attempted it. The first to turn his attention in this direction seems to have been Governor Winthrop, of Connecticut, who, from 1650 to 1660, engaged at intervals in examining the metalliferous indications of the Connecticut Valley, in the vicinity of Haddam and Middletown. There is no reason to suppose that any actual mining was ever executed by him.

About the same time, the Jesuit Fathers were exploring the Great Lakes of the far distant Northwest, and in the relations of their journeys for 1659 and 1660, we find the first mention of the copper of Lake Superior, which, occurring as it does in the native state, and not unfrequently found in loose masses along the shore of the Lake, was well known to the Indians, and could not escape the notice of so observing travellers as were Allouez and Marquette. Long previous to their time, at a period of which no record remains, this region had been the scene of extensive mining operations. The copper-bearing rocks had been explored throughout their whole extent, and even on the almost inaccessible island of Isle Royale, and mining excavations had been made at very many places. There is no reason to suppose that these workings were known to the Indians at the time of the first visits of the Jesuits; on the contrary, there seems to have been no tradition of them remaining, and the appearance of the excavations indicates beyond a doubt that they were made long before that period.

Just at the commencement of the eighteenth century, Le Sueur explored the Mississippi as far up as the St. Peter's River, expressly with the view of making discoveries of the metals. The lead deposits of that region did not escape his notice, but his real discoveries are so mixed up in the records of his journey with pretended ones, that the whole account has an almost fabulous air. Although he returned with a cargo of supposed copper ore, of which he fancied he had discovered an immense mountain, he must afterwards have recovered from his delusion, if such it was, since his voyage led to no farther explorations.

About the same time that Le Sueur was thus employed, a great step was taken in New England by the erection of an iron furnace, which was probably the first one on the American Continent. A business was thus commenced which was destined to increase with the growth of the country, and to be a greater aid to its development than any gold or silver mines could have been.

In 1709, the first chartered mining company in the United States came into existence, and the mining of copper was commenced, at Simsbury in Connecticut, where it appears probable that workings were carried on for several years, and some ore obtained. The success at this point seems to have led to the discovery of the same ores in a similar geological position in New Jersey ten years later, as it appears that the Schuyler Mine was discovered in 1719, and that considerable ore was taken from it and sent to England.

In 1719 and 1720, the attention of the French was directed to the Mississippi Valley, and the assumed existence of the precious metals in that region was made the basis of one of the most extensive and wildest schemes of speculation ever started. Both De Lochon and Renault, accompanied by a numerous corps of miners and mineralogists, explored the country near the confluence of the Mississippi and Missouri for the precious metals; but their operations were attended with no practical results beyond the discovery of the lead deposits, which were worked to a very limited extent for a while, and then, on the bursting of the bubble in France, entirely abandoned.

During the 18th century, a number of mining operations were

undertaken and carried on in different parts of the country, apparently with but little success. Between 1750 and 1760, the New Jersey copper mines attracted considerable attention, and were wrought to some extent. In 1762, the cobalt mine in Chatham, Connecticut, was worked; and about the same time, the Southampton lead mine was opened. None of these enterprises appear to have been conducted with much vigor or success.

During the latter part of this century, the western lead region began to be of importance. While Louisiana was in the possession of the Spanish, some mines were opened and wrought, but in a very rude manner, the ore being taken out from mere pits, and smelted on log-heaps. In 1798, improved methods of mining and smelting were introduced, and the business grew rapidly into importance as soon as the territory was ceded to the United States, which event took place in 1803. In 1774, Dubuque had commenced operations in the Upper Mississippi mines, near the town which now bears his name; but it was not until half a century later that the region was opened to general settlement, and the lead business began to be developed. At about the same time, by a remarkable coincidence, the Spanish lead mines of the Sierra de Gador were opened, and produced immensely, so that the price of that metal became very much depressed.

During the early part of the 19th century, pieces of gold were occasionally found in North Carolina, and from 1824 on, the search for this metal was carried on there to some extent. In 1829 and '30, however, it became quite general throughout the Southern States, and no little excitement was raised on the subject. Thousands engaged in the business of gold-washing, and in 1833 and '34 the amount collected in Virginia, North Carolina, South Carolina, and Georgia, was about a million of dollars a year. It afterwards fell off to the half of that sum for a few years, but when the mines in the solid rock began to be worked, it rose again to more than it had been when the washings were most productive.

Up to this time, the attempts at regular mining had been few and far between. Excavations had been made in the rock, but no extensive, permanent, and productive mine had been opened. The lead region of the West was indeed producing largely, but such was the nature of the occurrence of the ores, that systematic working and enlarged plans were not indispensable to mining them with success. And in other parts of the United States, there had never been encouragement enough, or sufficient mining skill, to direct the investment of any considerable amount of capital into that channel.

But in the mean time, the importance of our coal and iron had begun to be appreciated, and their production was increasing with rapidity. In 1820, the first cargo of anthracite was sent to Philadelphia, and in 1847 our annual consumption had nearly reached 3,000,000 tons; while our production of iron was over 500,000 tons, placing us about on a level with France, and only inferior to Great Britain as an iron-manufacturing country.

The mineral resources of most of the different states had been under investigation at the hands of the various state geologists since 1830, and their researches had been of great aid in defining the boundaries of the different geological formations, and thus limiting and directing explorations. As many of the persons thus officially employed were not practically acquainted with the nature of metalliferous deposits, they naturally fell into some errors in describing and recommending them to notice as worthy of working. Still, it must not be supposed that the most practised mine inspector, let it be one even who has spent his whole life in comparing the phenomena of veins and watching their actual development, can always say from mere surface appearances whether there is a prospect of successful working. With regard to the ores of iron, it is quite true that this may be the case; with a sufficient knowledge of the subject, the development of that business need never be a matter of uncertainty, so far as the occurrence, quality, and quantity of the ores is concerned. But in the case of those metals which are worked in deep mines, and on veins, of whatever class, it is often impossible to say before considerable expenditure has been made, what are the chances of success. This is especially the case in a new and unproved region, where the peculiar phenomena of the veins have not been studied, and where the changes of the metalliferous deposits at various depths which are due to local causes have not been satisfactorily made out.

In 1844, however, a new region was opened to the miner upon Lake Superior, to which the name of mining district may properly be applied, since the veins are numerous, well-defined, concentrated within a limited space, and some of them rich in metallic contents. Thither miners and speculators bent their steps, and after a few years of wild excitement and hap-hazard investments, the business became firmly established, and was in most cases prudently conducted. About the same time that the first dividend was paid on any mine, other than those of coal and iron, within the United States, the discovery of the golden treasures of California gave an impetus not only to searching for this metal, but to mining enterprises of all kinds throughout the whole country. The Atlantic States were searched from one end to the other; many long-abandoned mines were taken up; the gold region of the Southern States suddenly became the scene of an unheard-of excitement, and the whole country seemed to swarm with the promoters of mining enterprises. This general movement towards the development of our mineral resources was undoubtedly based in part on a sincere belief in the richness of our mines, and the possibility of new discoveries, which should furnish an opportunity for profitable investment; but it must be acknowledged that a large part of the excitement which sprang up in 1852 and '53, on the subject of mines, was the mere blowing up of a prodigious bubble, a repetition of the old Lake Superior speculations of 1845 and '46, on a new and wider field, and on a greatly enlarged scale.

The facility with which the public allows itself to be deceived, in regard to everything connected with mining, is as remarkable, as the machinery by which the swindling speculation is organized and brought into successful operation is simple. The locality is selected, and visited by some very distinguished scientific geologist, who for a sufficient consideration will write a sufficiently flattering report, and demonstrate the absolute certainty of success. The value of the mine is fixed at an enormous sum, and divided into one or even two hundred thousand shares; the company is organized, and the stock brought into the market. Every means possible is then taken to inflate its value; fictitious sales of ore are announced; the most flattering reports are received from the mine, and published in all the newspapers; the President of the company, who, perhaps, had never seen a mine before in his life, and who may therefore be excused for mistaking iron for copper pyrites, or perhaps even for gold, visits the scene of action, and finds the surface literally "covered with stacks of ore;" a series of dividends is announced as about to be paid, or perhaps, even, the ore or metal from a neighboring mine is purchased with a part of the capital paid in, and sold, and a dividend declared "from the proceeds of the mine;" the whole machinery of fictitious sales of stock is put in motion, the stock rises, and the promoters of the enterprise benevolently allow the public to step in and share with them in the magnificent profits which are certain to accrue. As soon as a sufficient quantity of the stock has been thus disposed of, and the getters-up of the scheme have pocketed the proceeds of their skilful manœuvring, the natural results follow: the stock, no longer artificially kept up, begins to droop; one after another the deceptions which have been practised become suspected; the unfortunate holders rush to dispose of their shares, but it is too late. The property which a few days before was quoted at hundreds of thousands can now hardly be given away; the unfortunate victims having nothing left as the tangible evidence of the brilliant dividends promised but the elegantly engraved stock certificates, and the equally valuable reports by which they were deluded.

And yet the mine, thus made the object of speculation, and perhaps abandoned in disgust, may be really of value, and capable of being worked so as to pay a moderate profit on the capital actually and judiciously invested in its development. But the idea was given out in the beginning of the enterprise that it could be made profitable at once, and because this has not been the case, the holders of the stock lose all confidence, and refuse to furnish the capital, without which hardly any mine, however rich it may be, can be put into a condition in which it can for any length of time be worked with profit. The system which prevails in this country of chartered companies with a large number of shares, seems especially adapted to make the mining business, which contains so much of the lottery element of uncertainty in it, a mere object of stock speculations.

The records of the last few years show almost without exception that companies with large fictitious capital, and an enormous number of shares, have been got up for the purpose of swindling the public, and not for *bona fide* mining purposes. It may be laid down as a universal rule, that the stockholders in a mining enterprise should be kept fully informed in regard to the expenditures and operations of the company. A frank and full publication is the only guarantee of sincerity and good faith. When these things are more generally understood, and the public refuses any longer to be victimized, we may expect to see a less noisy but far more effective development of our mineral resources than we have yet had.

It was with the hope of aiding, in some degree, in this development that this work was commenced, and is, not without some hesitation, given to the public. It seemed as if the time had come, when the history of our mining operations should be taken up, and our capabilities of production in this department be made the subject of a more comprehensive and general investigation than they had yet received. Fully sensible of the importance of the task which I was about to undertake, I have neglected no means to prepare myself to execute it with as much thoroughness as was consistent with the publication of the results within a reasonable time. All of our important mining regions on this side of the Rocky Mountains, and most of the prominent mines, have been visited by me, some of them repeatedly, within the last few years. Having also examined many of the most interesting European mines, I was enabled to make use of the materials derived from my own observations, and those published by others, to compare our own metalliferous deposits with those of Great Britain and the Continent of Europe. This seemed to me to constitute an important feature of a work of the kind projected, since it is chiefly by comparison of new mining districts with those which have been long worked, that light can be thrown on the former.

With these views, the arrangement of the work which has been adopted, is the one deemed best calculated to carry out the plan proposed by its title, namely, a description of our metallic wealth, or a detailed account of the resources of this country in the metals and their ores, and of the present state of the development of our mining interests, as compared with those of other countries. The first chapter is devoted to an explanation of the laws which characterize the deposits of the metals and their ores, and a brief description of the general methods followed in mining operations. This seemed necessary, in order that the technical terms used in the course of the work might not be unintelligible to the general reader, and especially that some clear ideas might be impressed, at the outset, as to the various modes of occurrence of the useful ores, and the importance in all mining operations of carefully recognizing these distinctions in their practical bearing.

In the succeeding chapters each metal is taken up, and treated, usually in separate sections, under the following heads: its mineralogical and geological occurrence; the distribution of its ores in foreign countries; the distribution of its ores in the United States. Under the first head it is not attempted to give a full mineralogical description of all the existing ores; for this information recourse may be had to J. D. Dana's standard treatise on Mineralogy. The principal ores have been noticed, and such information of economical importance given as seemed required in order to throw light on the succeeding sections. Under the head of the geological occurrence, the formations in which the ores of the metal treated of are chiefly found are enumerated, and the most important laws of the veins in their connection with the rocks in which they occur are stated.

In the second section of the chapter, a concise description

is given of the principal foreign districts where the metal under consideration is obtained, and occasionally a more minute account of mines which have an unusual importance from the amount of their produce, or the peculiar circumstances under which they are worked.

Finally, the mines within the limits of our own country are taken up, and described as completely as the materials which could be collected by personal examinations and from official or reliable published accounts would allow, or as was deemed essential to the plan of the work. Having no personal interest in any mining enterprise, beyond that of a sincere wisher that this branch of our national industry may be most speedily and effectually developed with the smallest waste of money, I have not hesitated to give my opinions in regard to the value of some of our metalliferous deposits, in the hope that when favorable they might lead to renewed vigor in their working, and that when unfavorable they might be of some avail in preventing useless expenditures. Whether my judgments will prove in most cases founded in truth (that they should be so in all could not be expected), time only can show. Of the imperfections and omissions inseparable from the first execution of a work somewhat new in its plan and object, no one can be more sensible than the author.

The statistical portion of the work has been carefully compiled from the best accessible data. All the foreign weights have been reduced to the English standard; for this purpose Mr. J. H. Alexander's excellent work has usually been followed as authority. The ton of 2240 lbs. has been used as the unit in the case of all the metals, except gold, silver, and mercury. This is the officially recognized weight of the ton in this country and in England, and it seems hardly worth while to change it, unless the whole absurd system of our weights and measures can be abolished, and one simple and convenient, as that of the French nation, adopted. In the tables of English sales of ores, the ton of 21 cwts. is used, in conformity with the usages of that country.

The sections of the mines are drawn upon the same scale, of 200 feet to the inch, throughout the work, so that a comparison of the amount of ground opened at any two different localities may be easily made. The shafts and levels are shaded with heavy oblique lines, and the ground stoped is indicated by the part in black.

THE

METALLIC WEALTH

OF

THE UNITED STATES.

CHAPTER I.

ON THE NATURE OF THE DEPOSITS OF THE METALS AND THEIR ORES, AND THE GENERAL PRINCIPLES ON WHICH MINING IS CONDUCTED.

Before entering upon a description of the mode of occurrence and geological position of the various metals and their ores, taken singly, it will be well to establish the meaning of some of the more important terms used in treating of subjects of this kind, and to lay down some general principles which the results of mining-experience have shown to be applicable to mineral and metallic deposits. The importance of this preliminary step will be acknowledged, when it is considered that these principles are derived from the combination and classification of facts observed in mining operations all over the world, and that they form the basis on which future enterprises of this kind should be organized. The results which have been obtained in this branch of applied science are drawn from a great variety of sources; and when we reflect on the expense and well-known uncertainty attendant on mining, it will be evident that, amid the multitude of isolated facts, to which every day is adding, some general principles should be recognized which may be relied on as a thread to guide one through the labyrinth of uncertainty.

Great as is the variety of forms under which the metalliferous deposits of different regions appear, and difficult as it might seem, at first sight, to discern any fixed laws in their development, yet a general consideration of their structure will justify the following classification.

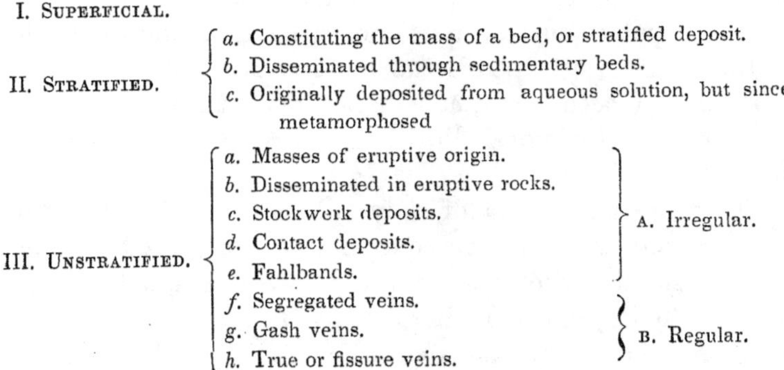

I. Superficial.

II. Stratified.
- *a.* Constituting the mass of a bed, or stratified deposit.
- *b.* Disseminated through sedimentary beds.
- *c.* Originally deposited from aqueous solution, but since metamorphosed

III. Unstratified.
- A. Irregular.
  - *a.* Masses of eruptive origin.
  - *b.* Disseminated in eruptive rocks.
  - *c.* Stockwerk deposits.
  - *d.* Contact deposits.
  - *e.* Fahlbands.
- B. Regular.
  - *f.* Segregated veins.
  - *g.* Gash veins.
  - *h.* True or fissure veins.

I. Superficial Deposits.—These comprise all deposits of the metals and their ores, which lie loose upon the surface, or intermingled with the superficial formations, the drift, and the alluvial. They are made up of particles and rounded fragments which have originally formed a part of some bed or vein, and from which they have been detached by various causes similar to those which have produced the great masses of clay, sand, and gravel, which almost everywhere cover the rocks in place. Such metalliferous deposits are usually called "washings," or "stream-works." Most of the gold, and all the platina of commerce, is obtained in this way, as also some of the tin. Their discovery and exploitation is hardly to be classed under the head of mining operations, although they furnish large amounts of the most valuable metals. Lying so near the surface, they can be reached by excavations of the most simple character. The dissemination of the metallic substances through the loose materials in which they are embedded is the result of mechanical causes, especially of currents of water, which have torn them from their original bed and scattered them over the surface.

II. Stratified Mineral Deposits.—This class of deposits comprehends those masses of ore which are included within rocks of sedimentary origin, and which are in every way identical in their epoch and mode of formation with the strata between which they are accumulated. This is not, however, the mode of occurrence of the more valuable metals. Of all the metalliferous ores, those of iron and manganese only occur in stratified deposits on an extensive scale. The beds of argillaceous ironstone, intercalated among the shales of the coal series, furnish the most striking illustration of this mode of occurrence, as will be fully set forth under the head of the Geological Position and Mode of Occurrence of the Ores of Iron.

There are also strata of slaty and other rocks which are impregnated with ore, or contain it disseminated through them in small particles, so thoroughly intermingled with the rock as to be hardly distinguishable from it. When such deposits occur on an extensive scale, and when the ore forms a persistent portion of the stratum, they become an important source of mineral wealth; although the percentage of ore is usually very small, yet from their great extent, they may be capable of being worked with profit. Such are the cupriferous schists and sandstones of the Mansfeld district in Prussia, and the similar deposits of the government of Perm in Russia. Further on, under the head of Copper, a description of these interesting mining regions will be given.

There is a class of metalliferous deposits which belong to rocks originally stratified, and of aqueous origin, but which have, since their deposition, been metamorphosed so as to have assumed in a great degree the appearance of igneous masses, and which themselves can with difficulty be recognized as possessing distinctly the character of either the stratified or unstratified groups. These are developed in the metamorphic strata of the palæozoic system, but especially in the azoic rocks, as in Sweden and in Northern New York, where the immense deposits of iron ore furnish examples of the class in question. They invariably coincide in their line of greatest linear development with the line of strike of the associated rocks, and are exceedingly irregular in their dimensions, expanding to a great width, and then contracting within very narrow limits. They usually also agree in dip with the planes of the rock enclosing them. Such masses of ore seem to have resulted originally from aqueous deposition, although the materials may in some instances have been derived from the abrasion and destruction of purely igneous masses. It is frequently supposed that these lenticular, or flattened cylinder-shaped deposits have been forced up in a plastic state from beneath, having been driven in between the strata like a wedge, separating them along the line of least resistance. This theory, however, seems hardly tenable, when we consider the great extent of such masses, and the absence of rubbed and polished surfaces along the line of contact of the ore and the adjacent rock. Moreover, such masses are sometimes found thinning out beneath, as if they had occupied a previously-existing depression in the strata, before they had been elevated from their original horizontal position. When such an upheaval of metallic matter has taken place from beneath, we find the result of mechanical displacement and chemical action in the surrounding beds to be most strongly marked. The azoic period having been one of long-continued and violent mechanical action, there is no reason to doubt that many of the masses of which it is made up may have been derived from the ruins of previously-formed rocks of the same age, both sedimentary and igneous. Such abraded material might readily have been swept into the depressions of the surface by strong currents, such as must have existed in those times, and would naturally have a very irregular thickness.

III. Unstratified Deposits.—This class comprehends most of the deposits of the ores of the metals, excepting some of those of iron and manganese, as noticed above. Their character is exceedingly varied, and the phenomena which they exhibit frequently of the most complex kind.

In the first place, they may be divided into two great groups, *irregular and regular deposits.*

Irregular Deposits.—These include first, igneous eruptive

masses, forming rock aggregations in every respect similar in form and mode of occurrence to the non-metalliferous rocks with which they are associated. This class of deposits consists chiefly of the immense masses of specular and magnetic oxides of iron, so strikingly developed in the azoic system, where they exist in connection with igneous eruptive rocks, and apparently under the same conditions and with the same mode of occurrence. In Europe, the Elba iron ores may be taken as a type of this class of ore deposits; in this country the Iron Mountain of Missouri, and the even more extensive iron-ridges of Lake Superior, furnish striking examples of a metalliferous ore existing in such quantity and under such circumstances as to identify it geologically with the adjacent rock-formations and to cause it to be classed with them, as having originated under the same circumstances. Such bodies of ore usually form ridges parallel with the lines of elevation of the rocky strata in which they occur, or they may take the form of dome-shaped masses which have lifted up the strata on every side, usually filling the fractured envelop with metallic emanations and showing an intense metamorphic action, in the formation of numerous mineral substances not elsewhere found occurring in the formation. As might reasonably be expected, such ore-masses are found developed on the largest scale in the azoic system, having been formed at a period in the geological history of the earth when its crust was thinnest and most easily fractured. If there have been eruptions of metalliferous masses during the later geological periods, they have taken place usually in a region of volcanic fires, or on deeply-seated fissures which have remained open until a late epoch.

As far as the practical application of mining science to this class of deposits is concerned, it is the most simple, for they are developed on so grand a scale that there is no question of exhausting them, and when worked, it is generally by excavations of the nature of the quarry rather than of the mine. In that part of this work which treats of the geological occurrence of the ores of iron, numerous examples will be given of this form of metalliferous deposit, which is almost exclusively confined to that class of metalliferous products.

Closely allied to the last-mentioned class is that of the metals and ores disseminated in the eruptive rocks, when the quantity of the metalliferous substance has not been sufficient to form a special deposit, or when the nature of the circumstances under which it was formed has not permitted its concentration. This is the mode of occurrence of platina and its associated metals, which are disseminated through eruptive trappean or serpentine rocks in fine particles, and rarely masses of a few pounds in weight; but which have never, thus far, been found in regular veins or in any concentrated form of deposit. Serpentine is frequently accompanied by chromic iron, which is disseminated through it in the same way, but much more abundantly. The trappean rocks contain frequently so large a proportion of the magnetic oxide of iron, scattered through the mass, that it may be considered one of the constituent minerals of the rock. Thus, in the mine of Taberg, in Sweden, the trap is so filled with bunches and strings of magnetic iron ore that it can be worked to advantage.

But in general, this class of deposits are of little value, unless nature has performed the operation of separating the metallic contents on a grand scale, by a gradual disintegration and removal of the rocky substance, and the consequent concentration of the much heavier metal or ore. This is the origin of all the platina-washings, which metal was derived from the abrasion of a rock containing it in so small a proportion, that it could never have been separated, by artificial means, with profit.

The tin mines of Europe are, many of them, worked in deposits analogous in their mode of occurrence to those now under consideration. For instance, at Geyer, in Saxony, the oxide of tin is found in small thread-like veins, and disseminated in fine particles through an eruptive dome-shaped mass of granite. It is from an original source resembling this, that the tin of the washings or stream-works of England and the East Indies has been in part derived. Oxide of tin is sometimes disseminated in invisible particles through the granitic rocks, and would probably be frequently found if carefully searched for. I have myself found 0·13 per cent. of this substance in a feldspar from the Riesengebirge, in Silesia, in which not a particle of it was visible to the eye.

Closely allied and passing into the last-mentioned mode of occurrence is that called by the Germans a "Stockwerk," a term which has been adopted into French and English, although not always correctly applied. The stockwerk consists of a series of small veins, interlacing with each other and ramifying through a certain portion of the rock. The ore is not usually entirely confined to the veins or branches, but exists in the adjacent rock in considerable quantity.

Fig. 1.

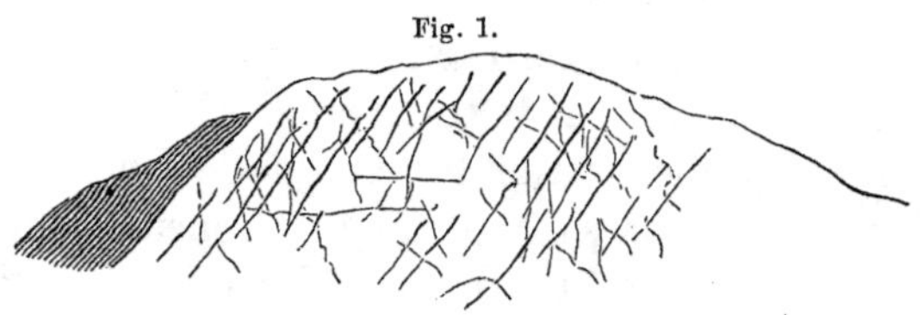

The annexed section (Fig. 1) represents such a network of veins, and the manner in which they may be imagined to traverse the rock.

Such reticulations of mineral matter are frequently called in England, *floors*, and in Germany "Trümerstocke," or "Stockwerke," in allusion to the fact that they are worked in different stages or stories, one above the other, owing to their great extent. Such deposits are usually proportionally poor in metallic contents, in an average of the whole quantity of rock which has to be worked.

The celebrated Carclase tin mine, near St. Austell, is a striking example of the stockwerk form of deposit. It was worked, until recently, in a decomposed feldspathic granite, as an open quarry. The ore occurs in a network of small veins interlacing with each other, the whole of which, together with the rock included between them, had to be removed and the ore afterwards separated by hand-picking, stamping, and washing.

There are other stockwerk mines in Cornwall, most of which have now been worked out and abandoned. The Altenberg stockwerk, in Saxony, furnishes another good example of this mode of occurrence. The rock is a grayish, silicious, feldspathic porphyry. The stanniferous mass is about 1400 feet long, and 900 wide, and like that described above, consists of a great number of small veins, which cross each other in every direction. Those which run nearly east and west, are usually the widest and richest in ore. The enclosing rock is also rich enough to be stamped and washed for a distance of several feet from the network of veins.

In the phenomena of *contact deposits*, *segregated veins*, and *true veins*, there is a gradual passage from one class of forms to the other, so that it is not always easy to say to which of these divisions a deposit of ore may belong. In the class of contact deposits, the ore is found concentrated between two formations of dissimilar geological and mineralogical character. Where the strata have been uplifted and metamorphosed by a central eruptive mass, not unfrequently a band of metalliferous ore, of irregular thickness, will be found extending along the line of contact of the eruptive with the metamorphosed rock. Or, if not always exactly upon the dividing line between the two formations, it may be found perhaps at an inconsiderable distance, and preserving a general parallelism with it. The annexed figure (2), illustrates, by an ideal section, this mode of occurrence.

Fig. 2.

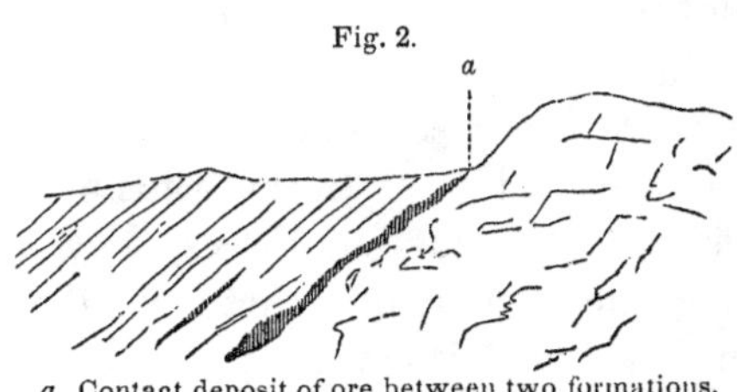

a. Contact deposit of ore between two formations.

In the same manner, beds of ore may be found between two successive overflows of an igneous rock, or metallic substances may occur disseminated through that portion of two beds of rock which is adjacent to the plane of separation between them; in such a case, the mineral masses on each side of the ore-deposit may have the same geological position, but will usually be found to differ in mineralogical character. Farther on, interesting examples of this mode of occurrence in the Lake Superior region, will be given. The iron ores of the Harz have a similar position in relation to the eruptive rocks of that district, since they follow the contact-planes of these igneous masses with the uplifted slates. The deposits of ore pass gradually into the trappean rock, showing that there has been a concentration of metallic matter at the junction of two dissimilar formations. The same phenomena exhibit themselves in the Vosges, near Framont, where masses of specular iron surround a central nucleus of quartzose porphyry, which has tilted up the stratified rocks, while the ore has insinuated itself into all the fractures and cavities thus produced, lining them with beautiful crystals and penetrating even into the solid rock itself. At this locality, and others resembling it, the action may be considered to have been principally of an igneous character, the upheaval of the strata opening a passage or chimney for the passage of metallic vapors which have become condensed in the fissures thus formed. In other cases, the phenomena are more satisfactorily explained, by supposing a segregation of the mineral substance along the line of junction of two rocks of dissimilar character, under the influence of electro-chemical agencies. The copper ores of Monte Catini, in Tuscany, belong, according to Burat, to the class of contact deposits, being developed along the line of outcrop of the gabbro, a rock resulting from the metamorphic action of serpentine upon the strata of the cretaceous formation.

Deposits of the character thus indicated are exceedingly irregular in their outline, and are often so combined with each other as to present the features of two or three forms at once, but, in general, in their essential characteristics, they are allied to the rocks in which they occur, in structure and composition. They are often developed at certain points on so large a scale as to be of great importance, but they have not that persistence in depth which is the characteristic of true veins. The ores which they furnish are less crystalline in their texture, and are often quite compact. The gangues are also less distinguishable from the adjoining rocks, and frequently there is no proper veinstone at all. In every mining district, one or the other of these modes of occurrence usually predominates and gives a character to the workings; if the deposits are of the irregular class, they are usually wrought with less skill and capital, although frequently on a very extensive scale; if, on the other hand, they are regular in their character, they require an enlightened system of working, and may be relied on as permanent sources of metallic wealth. Of this description are the mines of Cornwall, Saxony, and the Harz, which have been for hundreds of years the classic regions of mining, and in which the theory and practice of this branch of applied geology have been carried to the highest perfection.

Intermediate between the regular and irregular deposits of the metalliferous ores, are those forms of occurrence designated by the German word "Fahlband." These are more strikingly exhibited in Norway than elsewhere, although the same character has been recognized in other regions on a less extensive scale; and being in some respects peculiar, this mode of occurrence should not be overlooked in a review of the forms of metalliferous deposits.

The "Fahlbands" (German, Fahlband; plural, Fahlbänder), as developed in the district of the Kongsberg silver mines, consist of parallel belts of rock of very considerable length and breadth, which are impregnated with the sulphurets of iron, copper, and zinc, together with a little lead and silver, disseminated through the rock in such fine particles as to be hardly visible, and only to be recognized by their tendency to decompose, and thus to give to the rock in which they are contained a peculiar rotten and disintegrated appearance at the surface; hence the name "Fahlband," or, *rotten belt,* the word *fahl* being a corruption of *faul,* the miner's term for a rotten or decomposed rock.

In general these ore-bearing belts are irregular in their dimensions, although constantly preserving a certain degree of parallelism with each other. They may be traced in the Kongsberg silver-mining district for several miles, and the greatest breadth of any one is about a thousand feet. The quantity of ore contained in them is usually too small to be worth working; but occasionally it is sufficiently concentrated to become the object of mining enterprise. There are seven of these Fahlbands in the vicinity of Kongsberg, and they are parallel in strike and inclination with the gneissoid and schistose strata in which they occur, and have the same local character in relation to disturbances of stratification, schistose structure, and other external forms.

The fahlbands, however, are themselves traversed by fissure-veins bearing argentiferous ores, and the results of extensive mining have shown that these veins are only productive when they intersect the fahlbands, demonstrating that the impregnation of the lode with mineral matter was entirely dependent on the nature of the adjacent rock, and furnishing sufficient evidence that the metalliferous particles in the veins were originally derived from the fahlbands, and probably concentrated there by electro-chemical action.

This mode of occurrence will be recognized as peculiar in its character, being a combination of two distinct forms of metalliferous deposits; but the fahlbands themselves could hardly be considered of much importance, were it not for their enriching action on the lodes which traverse them. The same fact is observed with regard to the enriching of fissure-veins, when they traverse different beds of rock, even in cases where no perceptible metalliferous particles can be observed in any of them, and when the fahlband structure may be supposed to be wholly absent. In such cases, it is not unreasonable to suppose that a chemical examination of those beds, in which a lode shows itself as better filled with ore than elsewhere, might reveal the existence of metalliferous particles the presence of which had been previously unsuspected, because they were too finely disseminated through the rock, or were not of a nature to be easily decomposed, and so failed to give a marked external character to the stratum in which they occurred. The fahlbands may be considered as approaching nearest to the class of segregated veins, which comes next in the classification of metalliferous deposits given above, and to these, in connection with the other modes of occurrence to which the term "vein" is applied, we now turn our attention.

Regular Deposits.—Segregated Veins, Gash-veins, Fissure-veins.—The line of demarcation between the three forms of veins indicated above cannot always be easily drawn, although in most cases the difference is very apparent, and in all is of great importance in judging of the value of a metalliferous deposit. But in some instances, there is so gradual a passage from one form to the other, that surface examinations are not sufficient to enable one to decide the question of the actual existence of a fissure or true vein, and it is only by the indication obtained at some depth below the surface, that the vein can be placed in its proper class.

By a vein, as a geological and mining term, in general, is understood *an aggregation of mineral matter of indefinite length and breadth and comparatively small thickness, differing in character from, and posterior in formation to, the rocks which enclose it.*

Werner, the great Saxon geologist, defined veins as "mineral repositories of a flat or tabular shape, which traverse the strata without regard to stratification, having the appearance of rents or fissures formed in the rocks, and afterwards filled up with mineral matter differing more or less from the rocks themselves." This definition would exclude many veins which do not traverse, but run parallel with the strata, and others which occur in unstratified rocks. The definition given above, of course, includes veins of mineral matter not metalliferous, which are of frequent occurrence; but, of course, of no importance in the present connection.

Weissenbach, in a paper published in Cotta's "Contributions to the Knowledge of Mineral Veins," which is especially devoted to the subject of the metalliferous lodes of Saxony, has given a general classification of all the modes of occurrence which can be brought under the head of veins. He divides them as follows:—

1. *Veins of Sedimentary Origin.*—Where a fissure has been filled from above by deposition of mineral matter, in a manner similar to that in which the sedimentary rocks themselves have been formed. The origin of these is exceedingly simple, since it is evident that in the process of deposition of any of the stratified masses, as of sandstone or limestone for instance, if this operation were going on in a region where open fissures existed in the subjacent rock, they would be filled by the sedimentary matter, which would assume a stratified appearance in the fissure exactly as above it. The matter thus deposited would not properly be a metalliferous ore, since this class of products have a chemical and not a mechanical origin. This is perfectly evident, and it cannot

fail to strike one with surprise that a theory so inconsistent with facts should have been adopted by Werner to account for the formation of metalliferous veins. The whole class of veins of sedimentary origin is of little importance.

2. *Veins of Attrition.*—Fissures filled with matter introduced by purely mechanical means, such as by fragments of the wall-rock falling from above, or produced by friction of their sides against each other. These phenomena are exhibited in many ore-bearing veins, in the formation of which a twofold action was concerned.

3. *Veins of Infiltration, or Stalactitic Veins.*—These result from the filling of fissures by incrustation of the sides with calcareous matter deposited from aqueous solution, in the manner of the stalactites so common in caverns.

4. *Plutonic Veins.*—Fissures filled with mineral matter identical with that of rock formations, or mountain masses, and supposed to have been introduced by injection, or pressure from beneath upwards, while in a plastic state. Such are the common granite veins in granite or slate.

5. *Segregated Veins.*

6. *Metalliferous Veins, proper.*

These two latter classes are included in the class of regular, unstratified mineral deposits, as previously shown in the table of mineral formations; the latter division including both gash and fissure veins; and they are the only ones which are of importance in the consideration of the metalliferous veins.

Segregated Veins.—Under this class of metalliferous deposits are included those vein-like masses which have a crystalline structure, or, at least, a gangue differing from the adjacent mass, but which do not seem to occupy a previously existing fissure in the rock, being so enveloped and limited on all sides within it as to show that the metalliferous and mineral substances of which they are made up could not have been introduced into their present position in any other way than by a gradual elimination of their component particles from the surrounding formation. This process seems to have been of a chemical nature, and one by which materials of similar character were collected together from all directions, or segregated, as it is termed. Of the conditions under which the adjacent rocks must have been when such an elimination of their metallic contents took place, we know little with certainty. We see, however, examples of segregation in masses of lava as they cool from a state of igneous fluidity, when crystals of the different mineral species found in such rocks are found to have crystallized out into distinct individuals, from what was before an apparently homogeneous paste. The same is true of granite and the trappean rocks. In the former, the single crystals sometimes attain a length of several feet. If circumstances cause the segregating crystals to imitate an elongated mass, we have at once the rudimentary form of a veinlike mass, which may continue to develop itself, and acquire considerable dimensions. A tendency to this separation of pure quartzose material will be noticed in almost all of the so-called metamorphic rocks, which frequently show bands of pure quartz, parallel with each other, and lying in the plane of stratification of the enclosing rock; and it should be noticed that quartz is almost universally the veinstone of ores occurring in this form of deposit.

These segregated veins differ from true veins in some important respects. In the first place, they usually lie parallel with the cleavage planes of the formation in which they occur, which is the gneissic and schistose portion of the metamorphic palæozoic rocks. The annexed wood-cut (Fig. 3), represents an ideal section of the form of deposit here under consideration. The ore-bearing mass may or may not appear at the surface, but its extent downwards in one plane is not to be relied upon as in the case of a fissure-vein, since the accumulation is liable, at any point, to be found thinning out in depth, and transferred to another plane as represented in the figure.

Fig. 3.

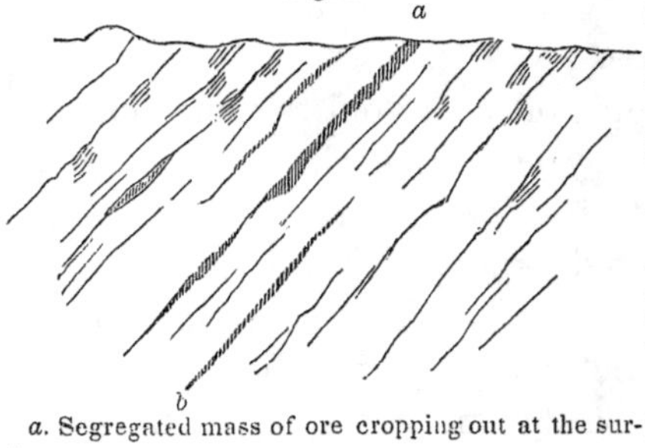

a

b

a. Segregated mass of ore cropping out at the surface.

b. Parallel layer not extending upwards so far.

A striking illustration of this form of deposit may be seen in the Rammelsberg, one of the most celebrated localities of the Harz, a section of which is given in the annexed cut (Fig. 4). The principal mass of ore (*a*, *c*), lies in the direction of the argillaceous slates of which the mountain is made up. At the depth of about four hundred feet, it sends off a branch (*b*), which dips at a considerably less angle than the slates themselves. The greatest thickness of the mass of ore is over one hundred and fifty feet; and its length is about nineteen hundred; but the dimensions decrease in depth, and at eight hundred feet its thickness is about twenty, and its length seven hundred and fifty feet; so that there can be little doubt that the mass will terminate entirely at a certain not very great depth. The ores of this singular mass are chiefly sulphurets of iron, zinc, lead, and copper, intimately blended together, and almost entirely destitute of gangue.

Fig. 4.

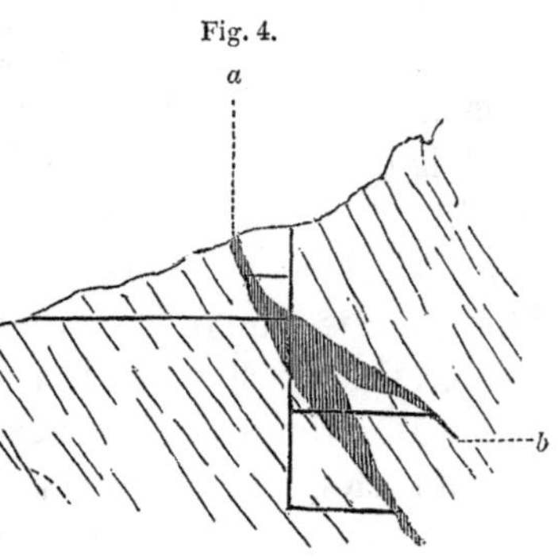

Section of the Rammelsberg.

The auriferous quartz-veins of most gold regions belong to this class of deposits. They consist of belts of quartzose matter, with sulphuret of iron, which near the surface is decomposed into a hydrated oxide, and contain gold disseminated through these substances, and sometimes in the adjoining rock, in fine particles, or, occasionally, large lumps. These belts run with the strata and dip with them, and in other respects exhibit the phenomena of segregated rather than of fissure-veins.

Practically, the most important feature of this class of deposits is that they cannot be depended on in depth as true veins; as they seem almost always to be richest near the surface, and frequently terminate altogether at no very considerable depth. Nor is the ore or metallic matter distributed through them with as much regularity as in the true veins, forming often a series of nests and pockets ranged in a general linear direction, and connected by mere threads of ore or barren veinstone.

Gash-veins.—This variety of mineral deposit holds an intermediate place between segregated and true veins. Like the latter, they occupy pre-existing fissures; but these are of limited extent, and not connected with any extensive movement of the rocky masses. They are usually confined to a single member of the formation in which they occur, terminating below, when a marked change in the lithological or mineralogical character of the rock takes place. The annexed cut (Fig. 5) represents, in an ideal section, the mode of occurrence to which the name of gash-veins is applied. The stratum *b*, included between *a* and *c*, cuts off the veins in *a* entirely, the fissures not having extended through that bed at all. Should circumstances favor, and the bed *c* resemble *a*, similar fissures may be found again below *b*, as represented in the figure, but it can by no means be asserted that they are the continuations of the identical fissures which were found in *a;* on the contrary, they are a new set, originating in and comprised in the bed *c*, as those above were in *a*. Lateral branches will usually be found in connection with the main fissures, which may or may not be nearly vertical, according to circumstances; but, whatever their position, the two sets of cracks will be nearly at right-angles with each other, and will possess the same character in regard to their mineral contents, although one set will generally predominate over the other greatly in extent.

Fig. 5.

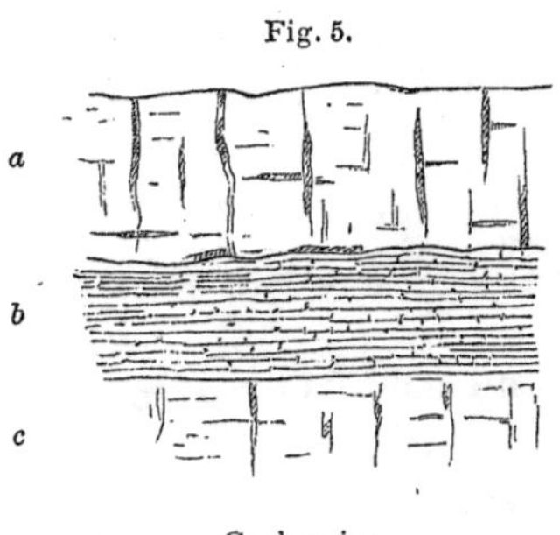

Gash-veins.

The origin of this class of fissures must, in all probability, be referred to the contraction of the rock caused by shrinkage, either while gradually undergoing consolidation, or from the effect of long exposure to a somewhat elevated temperature, after a cessation of which, certain strata might, on cooling, under peculiar circumstances of texture and thickness, be more liable to be fissured than others.

The filling of fissures thus originating with mineral substances may have taken place in various ways. Most of them would naturally be the recipients of the sedimentary matter in process of deposition, if they occurred in strata still accumulating. In a rock impregnated with metalliferous combinations, segregation would more readily take place from the walls of such cavities, and their contents would then have the characters of those of segregated veins. From true fissure-veins they differ especially in not showing as distinctly marked selvages, and in having a less crystalline and comby structure of the veinstone. They are especially to be found in the unmetamorphosed sedimentary rocks, where these have undergone so little change as not to have assumed a thoroughly crystalline texture, and have retained the original lines of stratification, which, in the fully altered rocks, are almost obliterated.

The difference between segregated and gash-veins may be sometimes hardly perceptible, but their origin is sufficiently distinct to justify their separation from each other; and there are mining regions where the peculiarities of one form or the other may be seen with great distinctness. The latter are still less reliable than the former, and are usually soon worked out in depth; but frequently their number makes up for a want of continuous extent in any one of them, so that a region where they abound may furnish, for a time, a large amount of ore.

True Veins.—A true vein may be defined as a fissure in the solid crust of the earth, of indefinite length or depth, which has been filled more or less perfectly with mineral substances; or, in other words, an aggregation of mineral matter, accompanied by metalliferous ores, within a crevice or fissure which had its origin in some deep-seated cause, and may be presumed to extend for an indefinite distance downwards.

True veins are almost universally admitted by geologists to have originated in "faults" or dislocations caused by great dynamical agencies connected with extensive movements of the earth's crust, and for this reason they are believed to extend indefinitely downwards; an assumption which is supported by facts, since no well-developed and defined vein has ever been found entirely terminating in depth. Gash-veins, on the other hand, as before remarked, occupying fissures which have resulted from shrinkage of the rock, cannot be expected to extend into strata of different character from that of the bed in which they originated.

Among all the forms in which the metalliferous ores occur, that of true veins is of much the greatest interest, since they are the principal repositories of the ores of the useful metals, and their exploitation is a matter of lasting importance, involving the employment of both skill and capital; hence their principal features require to be described somewhat in detail.

The linear extent of true veins is very various in different instances. Some of the longest known have been traced many miles; but, usually, even if they extend for so considerable a distance, they are not found to be impregnated with ore through the whole of their course. The longer the vein, as a general rule, the more likely it is to be, in some part of its course, rich in ores: thus, some of the great veins of Mexico, which have produced such enormous quantities of ore, have been followed for more than six miles, and have been opened and worked in a great number of places. The width of a vein is not necessarily in relation to its length; some, which are well-defined and traceable for a great distance longitudinally, are quite narrow. From the nature of their origin, their direction and dimensions must be somewhat irregular, since the two portions of the rock on each side of the original crack having been moved in relation to each other, the uneven sides of the fissure will have given rise to cavities of unequal width. That such motion has taken place, is abundantly proved, in numerous instances, by the actual displacement of strata or mineral masses once evidently continuous and now removed to a greater or less distance from each other both in a vertical and a horizontal direction. Such cracks frequently exist in the stratified rocks, and have not been filled up with mineral and metallic substances; in such cases, when there has been a perceptible vertical movement of the two sides in regard to each other, the break of continuity is called a "fault."

Fissures thus formed may have been the receptacles of mineral matter unconnected with any metalliferous ores, as in the case of dykes of trap or veins of granite, which are usually considered to have assumed their present position while in a plastic state. The filling of cracks in such cases was a process of short duration compared with the time required for the formation of veins containing metallic matter. In other instances there may be regular veinstone accumulated between the walls of a fissure, and yet ore be entirely wanting; but it is rare that this is the case, since in almost every vein in which mineral matter has slowly gathered, there have been metallic substances present in some part of its course, during some part of the time of its formation.

In regard to the occurrence of true metalliferous veins, there is no fact more striking than that they are rarely found singly, but rather in groups, often in a complicated network crowded into a comparatively narrow space. The great mining regions occupy but a small part of the earth's surface; while very extensive tracts are almost wholly destitute of metalliferous indications. Let any one take a map of Europe, and color upon it the best-known mining districts, which furnish the larger portion of the metals to commerce, and he will be astonished at the small space which they cover. The groups of veins of the Cornish and Saxon mining districts are so complicated in their number and variety of relations to each other, that centuries of working upon them have not yet fully developed even the more important facts with regard to them.

The larger portion of the vein-fissure is occupied almost invariably by the "gangue" or "veinstone." This is the earthy or non-metallic portion of the lode or vein, consisting of mineral substances, of which a few are of almost universal occurrence as associates of valuable ores. The principal one of these is quartz, which may be said to be almost never absent entirely from any vein. It occurs in a great variety of forms, usually more or less crystalline, sometimes beautifully so, especially in "vugs" or cavities of the vein. Next to quartz, carbonate of lime is most common, in the form of calcareous spar, sometimes compact, and sometimes crystallized, and often passing into brown-spar and dolomite. Fluor-spar and heavy spar are also minerals of frequent occurrence as veinstones. Sometimes these substances form the entire mass of the vein for some distance, either singly or together, and are completely destitute of metallic ores; but this is not usually the case.

The various minerals which make up the body of the lode are frequently arranged in a succession of plates parallel to its walls. Usually, these plates, or *combs*, as they are called, are made up of aggregations of crystalline matter, the separate crystals of which have their axes at right-angles to the wall of the lode, and are developed on the side turned towards its centre. The annexed figure (Fig. 6) represents part of a lode at Wheal Julia, near Binner Downs, in Cornwall.* In this, which furnishes a good example of a comby structure, we see a central predominating mass of crystalline quartz (*e*), on each side of which are seams of copper ore (*f*, *f*). This constitutes the main body of the lode. Between this central comb and the walls are two smaller combs of the same material, on each side, separated, in one instance, as seen at *c*, by a partition of earthy matter. Each comb consists of two corresponding portions, whose crystalline faces meet and interlock towards its centre.

Fig. 6.

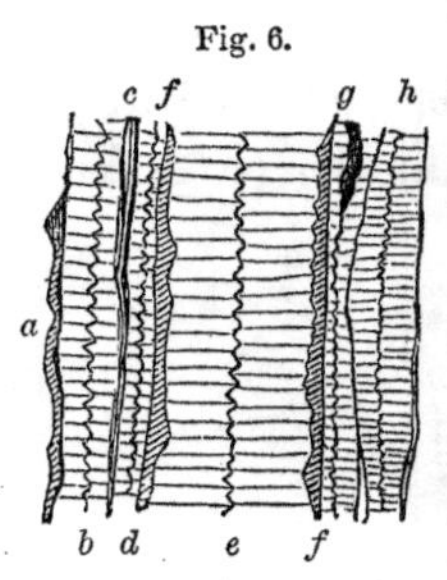

a, Bisulphuret of copper and sulphuret of zinc; *b*, comb of quartz; *c*, wall of indurated argillaceous matter; *d*, comb of quartz; *e*, larger comb of quartz, with blende and copper ore (*f*, *f*) on both sides; *g*, cavity, or vug, in another comb of quartz; *h*, more solid comb of quartz.

These appearances indicate a long-continued chemical action, occasionally broken off, and then renewed; the materials held in solution within the space of the vein, or gradually segregating from its walls, varying in character at different times. Sometimes a fissure may have been reopened, after having been once filled up with mineral matter, and

* De La Beche's Geology of Cornwall, p. 340.

may thus have afforded space for another deposition of veinstone. There are cases where this enlarging of the cavity of the vein seems to have been repeated, at intervals, several times in succession, the thickness of each comb indicating the width of the fissure at the time of its deposition, crystallization commencing on the walls as bases, and developing towards the centre, until the whole space was filled up.

In other comby lodes the width of the fissure remained the same, and a succession of deposits took place against its sides, the nature of the material varying at different periods, so that a section of the lode shows a series of various mineral substances arranged in corresponding parallel layers on each side of the centre. A beautiful instance of this mode of filling a vein-fissure may be seen in the annexed figure (Fig. 7),

Fig. 7.

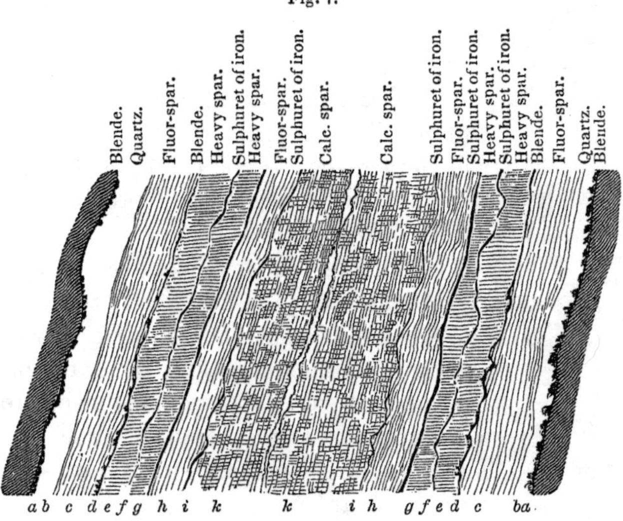

Fragment of the Drei Prinzen Spat Vein, near Freiberg.

from Weissenbach's "Illustrations of Remarkable Vein-phenomena," which represents a fragment of the "Drei Prinzen Spat" Vein, near Freiberg in Saxony. Next to the walls on each side is a crystallized deposit of blende (*a a*); to this succeed layers of quartz, followed by others of fluor-spar, sulphuret of iron and heavy spar, as indicated in the figure, each comb on one side having one exactly corresponding on the other, while the middle portion is occupied by crystallized calc. spar, with a cavity in the centre, the whole showing eleven symmetrical repetitions of six different mineral substances. Such perfect symmetry is, however, not often met with, and frequently the whole mass of the vein seems to have been formed by one uninterrupted process.

Every mineral district seems to have certain veinstones, as well as ores, in a measure peculiar to itself, and the locality of an ore may frequently be recognized by a simple inspection of a fragment of the accompanying gangue.

Besides the veinstones proper of a lode, there are frequently found enclosed in it fragments of the adjacent strata or wall-rock, which have been introduced mechanically. These may have fallen in, in some cases, from the surface, or in the formation of the fissure the rock may have been crushed and broken into fragments, which have afterwards been cemented together by the gangue, so as to form a brecciated vein. The fissure itself is frequently of a complex character, forming parallel branches, and sending out ramifications from the main line of fracture. Sometimes these strings are so numerous and irregular that the main fissure becomes lost, and can no longer be recognized by the miner. When a large mass of the wall-rock is thus enclosed between the branches of a vein, it is called by the miners a "horse," or the vein is said to "take a horse." The vein-fissure is sometimes abruptly contracted, so as to show nothing more than a simple crack; in this case it is said to be "nipped." The branches which leave the main lode are called "droppers," and when they concentrate or fall into it again they receive the name of "feeders." The direction and appearance of these off-shoots are closely watched by the miner in opening an untried vein, and from them he forms an opinion as to what parts of the lode are likely to prove richest in ore.

The mass of the veinstone is usually separated from the wall, by what are called "selvages" (French, *salbandes;* German, *saalbänder*). These are usually thin bands of clayey matter, and are of importance, since they prevent the adherence of the lode to the wall-rock, and of course facilitate its removal. The walls of the vein, themselves, are frequently smoothed and striated, as if there had been motion of the lode on its wall, accompanied by pressure: these polished surfaces are called "slickensides."

The annexed ideal section of a part of a vein (Fig. 8), will

Fig. 8.

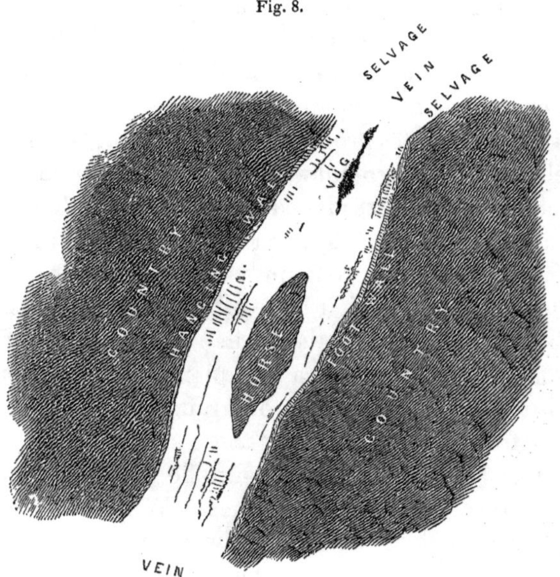

Transverse section of a vein.

serve to convey an idea of the various technical terms used in speaking of a vein, and its connection with the rock. Cornish miners having been first and most extensively employed in mining in this country, their terms have been generally adopted, as they are in England, from the fact that Cornwall is the most extensive mining district in that country.

It has already been noticed, that the metalliferous portion of a lode, or the really valuable ore, makes up, usually, but a small portion of its contents. The barren or worthless mineral substances extracted from the lode, and from the adjacent rock, in the process of removing the vein, are called *deads*, or *attle*. But few veins are sufficiently rich throughout the whole extent of the mine in which they are worked to allow of their being removed entirely. Frequently a large part of the lode is left standing, or unexcavated, while only the rich bunches of ore are taken out and raised to the surface. The occurrence of these rich deposits of ore in a vein is a matter of great uncertainty, and only to be judged of by the miner, after a careful observation of the district in which he is working. Some varieties of rock, or strata of peculiar mineralogical character, may be found, in certain districts, particularly rich in ore; and, in such cases, the right kind of "country" will be sought for by the miner, and the workings on the vein developed to as great an extent as possible in that rock. In other regions, when there are numerous parallel, or similarly situated veins, the *run of the courses of ore* (their situation in regard to the vein) having been determined in one case, the same rule may be found to apply to others. The differences in the character of two rocks, one of which may be highly favorable to the development of a rich lode, and the other just the contrary, are often not of a kind to be easily expressed in words. Long practice, and a well-trained eye, may be required, to decide at once which presents the most favorable appearance. Nor does the same rule seem to hold good in different districts, for the formation, which in one is productive, will sometimes be found barren in another. Hence miners, with certain fixed ideas, derived from the observation of a limited district, are apt to be led entirely astray, when they insist on applying the same rules to another mining region.

Since the crevices occupied by veins have resulted from dislocations of the rocks in which they occur, and since their formation has not been confined to any particular geological epoch, it follows that there may be, in any one formation, a variety of fissures of different ages and directions. This is found to be the case in many important mining regions, and the phenomena resulting from their influence on each other are often of great interest. The direction of a vein may sometimes be altered suddenly by a change in the course of the fissure, without being heaved out of its course, as it is termed, by the intersection of another. Thus, in the annexed cut (Fig. 9), which represents a transverse section of a vein at Holzappel, in Baden, where the vein coincides in dip with the strata, and has been shifted from one plane to another, a distance of twenty-five or thirty feet, without any dislocation having taken place in the rocks. The fissure still connected the two portions of the vein, but was so indistinctly marked that the level was driven through it, as shown in the figure, without its having been perceived; and it was only found again by sinking, when it was struck, having the same characters as before.

Fig. 9.

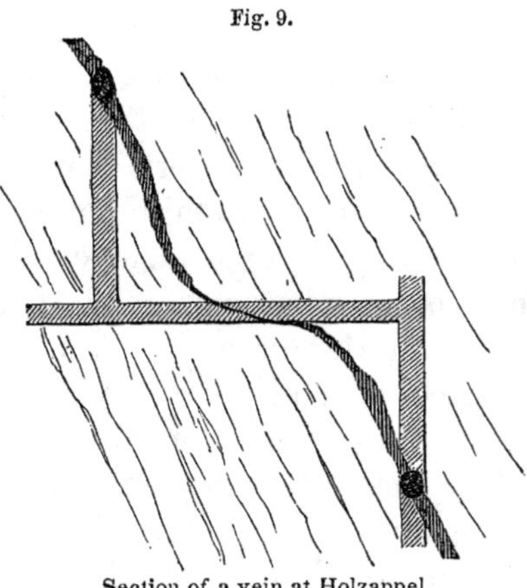

Section of a vein at Holzappel.

In a similar manner, a true fissure-vein may at some part of its course seem to coincide with the dip of the strata and actually send out branches which follow the lines of dip or cleavage; and yet the main fissure may pursue its course across the lines of bedding, although only cutting them at an acute angle, as represented in the section (Fig. 10). In such a case it may not be possible, without sinking upon the lode for some distance, to ascertain its real character as a true vein. The branches which are parallel with the stratification may be regarded as occupying fissures subordinate to the main one, which have been opened along the line of easiest fracture.

Fig. 10.

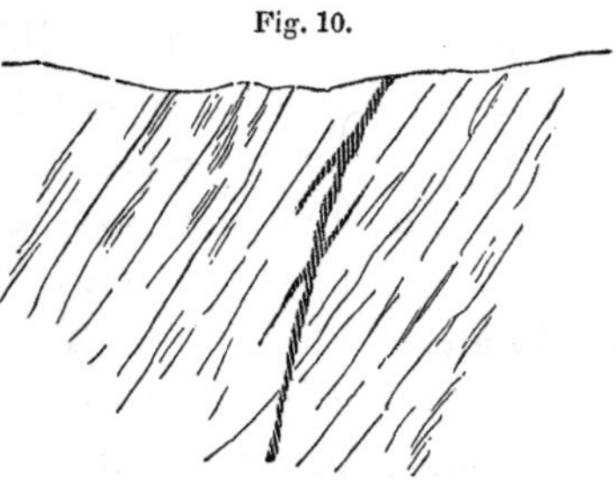

Veins, sending out branches which coincide with the bedding of the rock.

The derangements caused by veins of different ages and directions crossing are of great interest. If a vein traverses an older one it interrupts its continuity, and frequently "heaves" it to one side or the other. Two systems of veins thus crossing each other, may in their turn be both intersected by a third and still newer one, which may heave the other two. Instances of this kind are common in mining districts which have been subject to frequent dislocations. The newer vein intersecting an older one at a considerable angle is called a "cross-course," or a "contra-lode;" the latter name being usually applied to the intersecting vein if it contain valuable ores. In Cornwall, there are several systems of fissures, producing in many instances a complicated shifting of the veins. The general parallelism of veins of the same age is remarkable in many parts of that region, and generally in all districts where there are numerous fissures of different ages: it is also noticed that contemporaneous veins are not only parallel with each other, but that they usually contain the same varieties of ore associated with similar veinstones. In the Harz, there are two principal directions of fracture; the Samson, the most powerful lode of the district, cutting and heaving the Gnade Gottes and Bergmann's Trost, two other important veins. The most interesting and complicated phenomena of the intersection of numerous systems of veins are presented by the Freiberg district, in Saxony, where extended mining operations have been carried on for a great length of time, and have been studied by a skilful corps of mining engineers and professors attached to the School of Mines at that place. More than 900 different veins have been recognized within a space ten or eleven miles in length, by four or five in breadth. They are grouped by Weissenbach into four classes, according to the nature of their gangues; each of which is also in some measure characterized by certain ores, and by a constant direction.

In this country there have been no such complicated phenomena of veins of different ages observed. In the region to which most properly the term mining district may be applied, namely, that of Lake Superior, there are numerous veins, but they do not afford evidence of more than one epoch of formation. In any limited district, as for instance that of Keweenaw Point, the veins are nearly parallel with each other, and are not intersected by cross-courses which heave them from their course. In a few instances there seem to have been slips of the different beds of rock upon each other, which have shifted the veins at the junction for a few feet to one side or the other. In the Southern States the metalliferous deposits usually lie in the direction of the stratification, belonging to the class of segregated veins, and are not crossed by any fissure-veins. It will not therefore be necessary to go into any lengthened description of the particular appearances attendant on such intersections, especially as they differ much in different districts. In general, however, the crossing of two veins is wont to be considered as advantageous to the production of ore at that spot, and rich bunches of mineral are looked for by the miner at their intersection, especially if the angle at which they meet is a small one. Sometimes one lode impoverishes the other for some distance, carrying all the ore with it. As an instance of this appearance presented by two lodes intersecting each other, the annexed cut (Fig. 11) is given, representing one lode crossing another as sketched by Weissenbach.

Fig. 11.

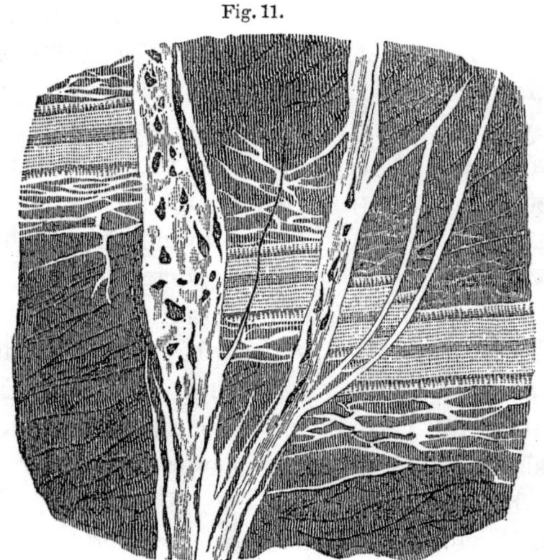

Ground plan of the intersection of veins in the Himmelfahrt mine, near Freiberg.

Theories of the Formation of Mineral Veins.—Although a full discussion of the various theories of the formation of metalliferous deposits would occupy much more space than can be given to the subject in the present work, yet it will not be out of place to notice briefly some of the most important theoretical considerations which the study of vein-phenomena suggests. The following are the principal heads under which the theories which have been proposed may be arranged.

1. The veins originated contemporaneously with the rock in which they are contained, and are, so to speak, a mere accidental phenomenon, not governed by any fixed laws of formation. This theory may be dismissed at once, as entirely at variance with all the facts, and as unworthy of consideration.

2. Veins have originated in the filling of fissures, by injection of metallic and mineral matter in a state of igneous fluidity from below. This is the theory usually adopted to account for the phenomena of the veins of so called igneous rocks, such as granite and trap; which, like modern lava, are supposed to have been once in a plastic or semi-fluid state, under the influence of a high temperature, and in such a condition to have invaded the superincumbent rocks, being forced into the crevices by upward pressure. However such a theory may adapt itself to the Plutonic veins, it cannot be considered as explaining the modes of formation of metalliferous lodes. It fails to account for, or rather is contradicted by, the often observed fact, that the character of the lode changes with the rock in which it is found, being rich in ore in one formation, and barren in another adjacent one. This could not be the case if the vein had been forced up through the strata, as the nature of the rocks through which it was raised could have had no influence on its contents, the action being but momentary and mechanical. Besides, if we consider the immense force which must have been required for such an upward motion as this theory supposes, it will be apparent that, had it really taken place, evidence of its existence must have remained in the widening out of fissures in depth, and in the shattered condition of their walls; while there would have been a constant tendency in the more valuable metalliferous substances, being heavier than the veinstone itself, to occupy the lowest position in the vein. Such phenomena, however, have only been observed in isolated cases, while usually the appearance of the walls and the distribution of the mineral matter and ore between them

in true metalliferous veins, is such as to make this hypothesis of their formation entirely untenable.

3. The theory of formation by sublimation, according to which vein-fissures were filled by the volatilization of metallic matter from the great centre of chemical action beneath, namely, the ignited interior of the earth. That such may have been the origin of some metalliferous deposits, and that this agency may have contributed in some degree to the filling of veins, cannot be denied. The fact of the volatility of some metallic combinations is well known, and can be observed at the present day in the products of volcanic ejections. Evidence of the same character is afforded, in some instances, by the position of metalliferous particles on the under side of crystals lining the walls of a lode; as, for instance, at Nagyàg, in Transylvania, where metallic arsenic is seen to have been sublimed and deposited on those faces of crystals of manganese-spar which were turned downwards. Specular iron is found sublimed into the fissures of volcanic craters, and sometimes carried to a considerable distance, and deposited. But these phenomena are of limited extent, and not by any means sufficient to account for the existence of the masses of ore and of earthy minerals filling the body of a large vein. Neither would such a theory account for the variation in the character of lodes in passing from one kind of rock to another, nor for the presence in them of substances not volatile in their nature, nor for any of the complicated phenomena exhibited by veins in their intersections with each other. Hence we must conceive that the agency of sublimation was of very secondary importance in the formation of regular metalliferous veins. In contact deposits, and some other irregular forms of occurrence, where the whole mass of a bed seems to have been impregnated equally throughout by metallic particles, as especially exemplified in some mercury mines, we can conceive of no theory more probable than that of the diffusion of the metallic matter through them by sublimation. Thus originated the extensive beds worked at Almaden, so rich in mercury; and they offer the most striking example which can be given of the class of deposits to which this theory may be applied.

4. The theory proposed by Werner, which may be called that of aqueous deposition, presupposes a chemical solution covering the region in which the veins are found, from which solution, by chemical precipitation from above downwards, the vein-matter was accumulated in the fissures existing in the rocks below. This theory is in direct opposition to that of igneous injection, since, according to its principles, the origin of the contents of veins was a superficial one, their introduction into the fissures from above instead of from below, and the action a chemical instead of a mechanical one. But, in the sense in which this mode of formation was understood by Werner, but little importance can be attached to it. If any such fluid holding metalliferous substances in solution had actually covered the surface, we can conceive of no reason why it should have deposited its contents in the fissures rather than on the surface adjacent; and we ought, in accordance with his ideas, to find every vein connected with a flat sheet of metalliferous ore somewhere along its course, at the place which the solution occupied in the series of formations at the time of the filling of the vein-fissure. Such, however, is not the case, at least in regard to true veins; although there may be a limited class of mineral deposits to which this theory will apply. Besides, if deposition in veins took place in this manner, we should expect to find more or less matter introduced at the same time, mechanically, and showing its origin by its stratified condition. There is nothing of this kind, however, observed in true veins. The deposits all took place in a direction parallel to the walls, and not horizontally, as they would have been, in part, under the circumstances required by this theory. There are many other reasons, equally conclusive, against the ideas of Werner; but it is not necessary to enter into them at length, since his theoretical views with regard to the origin of veins have ceased to have the weight which was once attached to them.

5. The theory of lateral secretion. The views which are at present most generally adopted, assume a somewhat complicated series of phenomena as concurring in the formation of mineral veins. It cannot be doubted that the process has been a complex one, and one requiring a long period of time

for its development. No one simple cause can be considered sufficient to account for all the facts, but the main idea is that of *lateral secretion*, or segregation of the mineral and metalliferous particles from the adjoining rocks in a state of chemical solution, and their deposition upon the sides of a previously-formed fissure under the influence of electro-chemical forces. This is the only theory which will account for the often noticed fact of the change in character of a lode in passing from one geological or mineralogical formation to another of a different character, and it seems to be more in accordance with the other phenomena of veins than any other one yet proposed.

The metallic appearance of the most common ores and their resemblance to certain well-known furnace products, are apt to lead to the supposition that they must be exclusively of igneous origin, especially in view of the fact that the mineralizing substances with which the metals are generally found combined are such as have the effect of rendering them more volatile. Nor can it be denied that some ores and metals do occur in rocks of undoubted igneous character in such a way that they must have had an origin contemporaneous with that of the mass in which they are disseminated. But these deposits do not belong to the class of regular veins, in general, but are to be classed, as already shown, among the irregular modes of occurrence. The metalliferous substances produced artificially, and found filling cracks and cavities in connection with furnaces, and in the chimneys of smelting-works, do not seem to possess that comby structure, or parallel arrangement so characteristic of true veins. If metalliferous particles were introduced into a vein-fissure by sublimation, they would naturally rise and be condensed together in accumulated masses, which might have a crystalline structure, but would be distributed in such a manner as to show their igneous origin.

On the other hand, we know that the earthy substances forming the gangue of metalliferous lodes, and usually far predominating in quantity over the ores themselves, are not the products of igneous action. The minerals which form the mass of the trappean rocks and of volcanic lavas, such as the feldspar family, hornblende, augite, olivine, and the like, are not the usual components of the veinstones, as we should suppose would have been the case, were the veins in which they occur the result of igneous action. On the contrary, we know that the usual gangue minerals, quartz, calcareous spar, and heavy spar, according to the researches of Bischof and other eminent chemists, cannot possibly have been introduced into vein-fissures by injection while in a state of igneous fluidity, or by sublimation. We know also that some of the ores accompanying these veinstones undergo decomposition at a high temperature; and many of the most common, such as the sulphurets of iron and lead, we see in constant process of formation in aqueous solutions, at the present time. The sulphuret of iron is one of the most common products, where ferruginous waters are brought into contact with decaying organic matter; and it is also one of the precipitates from hot springs. It has been shown by Murchison that the copper ores of the Permian strata of Russia must have originated in a similar way, since they are accumulated around and on the surface of the carbonized remains of the stems and branches of plants.

Fissures being opened far down into the heated interior of the earth, they would become filled with water charged with various kinds of metallic and mineral substances. In their passage through rocks of various characters these solutions would be variously acted on, and would, under hydrostatic pressure, penetrate more or less deep into the surrounding strata, according to their texture and other circumstances. If the water was acidiferous, it might thus combine directly with metalliferous particles, previously existing there. There can be no doubt that the contents of a vein-fissure have often acted chemically upon the adjacent rocks, since it is very frequently found that the walls have undergone a great change for a considerable distance from the lode; often the "country" is found decomposed and rotten, its texture being entirely destroyed; or it is impregnated with silicious matter, so as to be much harder and more quartzose than at some distance from the lode. These hardened silicified masses are called by the Cornish miners the *capels* of the lode.

Under the presupposed circumstances, it is evident that a fissure being filled with solutions thus originating, there would be ample reason why there should be deposits of ore within the limits of certain strata, in preference to others. Either those beds might themselves contain more metalliferous substances, which would enter into solution, to be again deposited within the walls of the fissure, or there might be something in their structure which would enable them to act with more efficiency in decomposing the solutions brought from other sources, by circulation of the column of water filling the cavity of the vein.

We cannot doubt that electro-chemical action is, and has been going on in mineral veins, and that the explanation of many of the more obscure facts with regard to the distribution of the ores within them is to be found in this all-pervading agency.

Mr. Robert Were Fox seems to have been the first to be impressed with the idea that electric currents might, "not only contribute to produce the extraordinary aggregation, and position of homogeneous minerals in veins," but that the action was still kept up in them, and he instituted a series of experiments in the Cornish mines, for the purpose of determining this fact. He found the mine-waters to be impregnated with a variety of salts, but in very different proportions in different parts of the same mine. He also found by numerous experiments that the reaction of these solutions on each other, and their contact with extensive surfaces of rocks of different characters, gave rise to electrical currents, the effect of which was to promote decomposition and deposition of various mineral substances, similar to those found occurring in veins. He thus was enabled to account for some of the most curious facts in relation to the position of the ores in the Cornish mines. Professor Reich, in the Freiberg mines, arrived at similar results, and Mr. Robert Hunt also made numerous experiments, determining the existence of galvano-electric currents in the veins of Cornwall, and producing chemical decomposition by their aid. He arrived at the conclusion that these currents were purely local in their nature, originating in the decompositions going on in the vein itself, and not connected with the great magnetic forces which sweep around the earth.

Whether or not it be allowed that these electro-chemical currents performed a very important part in the original formation of veins, it cannot be denied that in the chemical changes which have taken place in the upper portions of metalliferous lodes, and which are evidently still going on, they were and are conspicuous agents. These changes of character are well recognized in all mining districts. They are usually indicated on the surface by a discoloration of the lode by oxide of iron, and a general rotten appearance of the veinstone. The Germans call this the "iron hat" of the vein, the Cornish give to it the name of "gossan." In the cupriferous lodes of Cornwall, the normal ore, below the point to which decomposition has reached, is copper pyrites, or sulphuret of iron, and copper, and other sulphurets of copper. But near the surface, the metalliferous ores, instead of the sulphurets, are oxides and oxidized compounds, such as silicates, carbonates, and phosphates. Often a large portion or all of the copper near the surface has been removed, having been oxidized to sulphate of copper, and dissolved out and washed away, leaving the iron mixed with the quartzose veinstone, in the form of a hydrated oxide, or gossan. Similar appearances are presented by lodes containing lead ores, the sulphuret of lead, or galena, which is almost the only ore found in any considerable quantity at some depth, being, near the surface, converted into a great variety of oxidized combinations, of which the carbonate, sulphate, and phosphate are the most common.

In general, this zone of decomposition does not reach much below one hundred feet, but in some instances it penetrates to the depth of three hundred, and there are, on the other hand, veins in which no such chemical changes have taken place, the original sulphurets existing undecomposed almost at the very surface. The causes which have operated in accelerating and retarding these transformations are not quite clearly understood. Bischof has shown that heated steam has been one of the most powerful agents in the decomposition of the contents of veins, and that it has been most effectually aided by carbonic acid. In order to prove this, he subjected sulphuret of lead, while heated, to the influence of a current of steam, which reduced the galena to metallic lead, the sulphur passing off in the form of sulphydric and sulphurous acid gases. Sulphuret of silver is reduced in the same way, the resulting metal taking that dendritic form which it commonly has in its associations with the ores of lead and silver. The reduced metals thus set free, if they belong to the oxidizable class, are then ready to enter into combination with the various acids everywhere being liberated under electro-chemical agencies. No doubt the effect of atmospheric influence is to be largely taken into account also, since these changes are principally superficial only, compared with the total depth to which the ores are found to extend. The penetration of water charged with oxygen into the fissures and minute pores of the rocks must be in some degree essential to these processes. To give any idea of the complicated changes thus slowly taking place in many lodes, would require far more space than the object and scope of the present work could permit; it is sufficient here to have indicated some of the causes at work in the formation and transformation of mineral veins.*

The variations of the metalliferous portions of lodes at different stages are of the highest importance to the miner, not only in relation to the changes which the ores have undergone near the surface, but as viewed in connection with their persistence in mines worked to great depths. In this country we have hardly more than begun to open our mines, so that we have but little more than superficial indications to judge from; but in the mining districts of Europe, which have been wrought for centuries, there are abundant data accumulated to enable us to draw some important conclusions in regard to the termination of veins in depth. The distinction between true veins and all other forms of occurrence of ores must be carefully studied in reference to this question, or otherwise it will be difficult to obtain a clear

* For further information in regard to this class of subjects, the reader is referred to G. Bischof's "Lehrbuch der Chemischen und Physikalischen Geologie," which has been in course of publication since 1847.

conception of the truth. The facts may be briefly summed up thus :—

1. True fissure-veins are continuous in depth, and their metalliferous contents have not been found to be exhausted, or to have sensibly and permanently decreased, at any depth which has yet been attained by mining.

2. Segregated and gash-veins, and the irregular deposits of ore not included under the head of veins and not occurring in masses as part of the formation, cannot be depended on as persistent, and they generally thin out and disappear at a not inconsiderable depth; at the same time they are often richer for a certain distance, and contain larger accumulations of ore, than true veins, so that they may be worked for a considerable time with greater profit than these, although not to be considered as of the same permanent value. Of these assertions, many illustrations will be given in the course of this work, so that a clear idea may be formed of the distinction which is here intended to be drawn between deposits of a persistent and those of a temporary character.

THE GENERAL PRINCIPLES ON WHICH THE CONTENTS OF A VEIN OF MODERATE WIDTH ARE REMOVED.

In order to render the descriptions of mines and mining operations in the succeeding chapters more intelligible to those who have never given their attention to these subjects, it is proposed to add a few pages on the general principles of *exploitation;* by which term is understood the process by which ores and minerals of value are won from their natural position, often at great depths below the surface, and brought where they can be rendered available. Those who are not acquainted with the nature of such operations, and who have never had occasion to reason on the many circumstances which must be taken into consideration in laying them out, frequently suppose that a mine is a mere excavation in the ground without law or rule, and that this, which would seem to them to be the simplest method, would be also the best for such a work. Nothing can be farther from the truth; for to open a mine, if on a vein or regular deposit of ore, and work it as an excavation open to the air, or as an *open-cut*, is quite impossible, if it is necessary to go below the most moderate depth. And such is the case with metalliferous lodes, the superficial portion of which, for some distance down, is often less valuable than that below, and sometimes of no value at all. Moreover, the richest part of a vein is usually of limited extent, and its treasures can only be won by penetrating into the earth to great depths. To raise the ore from such a distance beneath the surface, requires powerful machinery which must be placed on firm ground directly over the vein. The surface-water and rain must be kept from running into the excavations, which could not be done were they open, throughout, to the day; and as the freeing of a mine from water is frequently one of the most expensive operations connected with its working, the greatest care should be taken to get rid of as much as possible in the simplest and least expensive manner. Again, it is necessary that mining should be carried on uninterruptedly by night and day, and in all weathers, since only a limited number of persons can be usefully employed at one time; also that a system of ventilation should be kept up; all of which would be impossible, were the excavations exposed to the weather, or the mine made to consist of a single open cavity.

The first operation in opening a mine is usually to remove the surface-water by driving an *adit-level* or horizontal gallery, which shall commence in an adjacent valley, at the lowest point not liable to inundation and conveniently accessible, and thus drain all that part of the proposed work which lies above. Figures 12 and 13 illustrate the position and object of the adit-level. In extensive mines large sums are frequently expended in thus getting rid of the water; sometimes a number of mines are drained by a simple adit of great length driven at their joint expense, or that of the state, when the work is of great magnitude. So long as the exca-

Fig. 12.

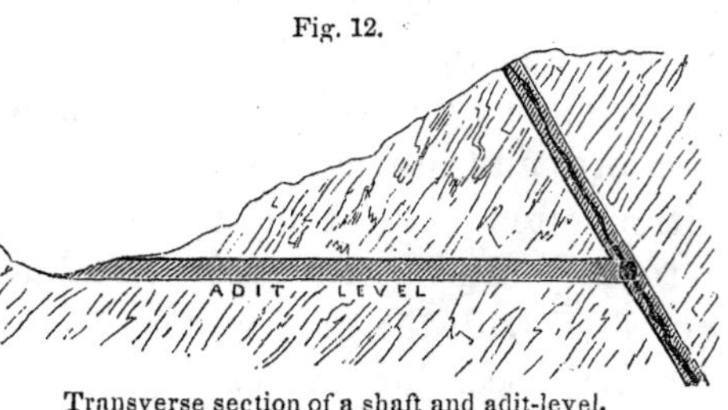

Transverse section of a shaft and adit-level.

vations in the mine are kept above the level of the adit, no machinery for pumping is required; and when extended in depth below it, the water is elevated by the pump or other contrivances employed, only up to that point; thus, in a very extensive mine, every foot through which a large body of water is to be raised, which can be saved, is of importance, and large sums of money may be economically expended in obtaining even a small reduction of the distance.

Frequently only a few feet of "back," as it is called, or of perpendicular height above the adit-level can be obtained; but, in other cases, by an adit planned with judgment, several hundred feet of drainage may be obtained, so that mining may be carried on for a long time without the expense of costly machinery for pumping. If possible, the adit should be driven on the vein, or by the side of it, so that its character and richness may be ascertained by breaking into it occasionally as the work progresses. A large number of the Lake Superior mines are thus favorably situated; and extensive drainage may be had at the same time that the vein is proved in depth.

If sufficient advantages present themselves for the location of machinery for washing and dressing the ores at the mouth of the adit-level, they may be brought out on a tram-road laid in the level, and the necessity of hoisting them to the surface by machinery may thus be obviated. If the adit-level cannot be excavated *on* the vein, it must be driven *to* it, as represented in Fig. 12, and the levels are then to be extended on it right and left. In a region where the veins run in a direction parallel with the range of elevations, it will generally be necessary to drive the adit-level through the "country;" but if the vein cross the ridges at an angle, it will also cross the nearest valleys on each side of the elevation, and may be worked from one side or the other. In the majority of cases, however, the adit-level is only used to remove the surface-water, but few mines being so favorably situated as to have a very large portion of their work above it.

As the necessity of ventilation would not allow the adit-level to be driven to an indefinite distance without some communication with the surface, so as to establish a current fitting up the machinery for pumping, tram-roads, and other requisites.

The dimensions of a shaft are variable, and dependent on the work to be done by its aid; when it is to be used for winding up the kibbles, or buckets of ore, and at the same time as an "engine-shaft," or the shaft in which the pumping machinery is placed, it should be from twelve to fourteen feet long, and is usually from six to eight feet in width. Shafts which have to be timbered up for the whole, or a portion, of their length, are square or rectangular; those which are built up with stonework are usually round. In this country, wood is universally used for this purpose, timber being usually cheap and abundant in our mining districts. In the Lake Superior region, the shafts only require to be supported for a few feet, or until solid rock is reached. The timbers are squared roughly, and laid upon each other, forming a substantial framework, which rests in a bed-place cut in the rock where it is sufficiently solid to obviate any danger of its giving way.

For a distance of fifty feet, or even one hundred, if desirable, the ore and rubbish may be raised to the surface by the simple windlass, worked by hand, on which a rope is so wound that one bucket or kibble descends while the other ascends. The same means are used in sinking a *winze*, or excavation from one level to another, not extended up to the open day. As soon, however, as the depth of the shaft becomes more considerable, it is necessary to resort to horse or steam-power for raising the ore. The common machine used for this purpose is called a whim; as usually constructed when worked by horse-power, it is represented in the annexed cut (Fig. 14). In the steam-whim, which is generally used when the shaft has a depth of more than two hundred feet, the cage or drum on which the rope or chain is wound is usually placed horizontally, instead of vertically as in the common horse-whim. Frequently, in opening a mine, before the excavations have reached any considerable depth, the pump-rod is attached to the whim, and worked by the same power which raises the ore.

The steam-engine is not usually applied to working a

mine, until there is good reason to suppose that it may become of permanent importance. For hoisting ores, no very great power being required, small horizontal high-pressure

Fig. 14.

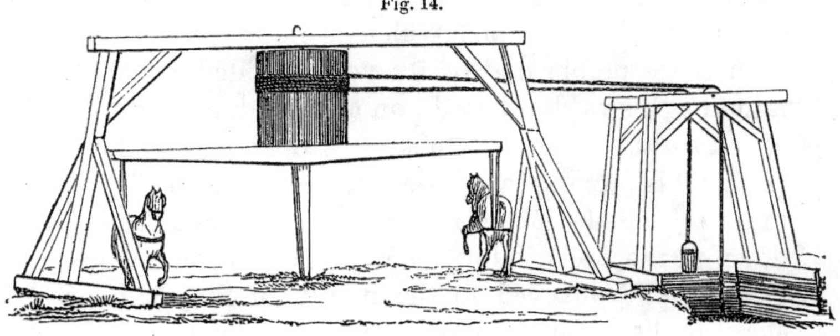

Horse-whim.

engines, working full steam, are usually employed. The machinery must admit of instantaneous reversal, and should be as simple as possible. Water-power, if at hand, can be advantageously applied to the same purpose. The pumping-engines in extensive mines are usually low-pressure ones, and every effort has been made in Cornwall and other mining regions to reduce the consumption of fuel to the smallest possible amount, by perfecting the form of the boiler and machinery.

Most mines, when worked to any considerable extent, have several shafts; two, at least, are almost necessary for ventilation. The distances from each other at which they are located depend on a variety of circumstances, such as the nature of the surface, the distribution of the ore in the vein, and the extent of the sett or length of the vein which can be worked. Two shafts should not be too near each other, if it can be avoided, as in that case the blocks of ground would be too short, causing a wasteful expenditure in opening the ground. If far apart, winzes must be sunk between them, in order to afford ventilation and the means of attacking the vein at a sufficient number of points. In general, three hundred feet distance between two shafts, in a vein of moderate width and richness, may be considered reasonable.

After the shaft has been sunk to a proper depth, if on the vein, drifts or levels are commenced from it in each direc-

Overhand stoping is almost universally practised in this country. The annexed section (Fig. 15) will illustrate this method. It represents a supposed longitudinal section, on the vein, of a shaft and two levels, between which stoping is going on. The rubbish which is necessarily blasted down with the valuable part of the vein, is piled up back of the miner, as his work proceeds, on a scaffolding of stout timbers, called *stulls*. At suitable distances, openings are left through the rubbish (one of which is indicated in the figure by the letter *m*) called *mills* or *passes*, through which the ore is shot down to the level below, to be loaded on to a wheelbarrow or car, and conveyed to the mouth of the drift, or to the shaft, to be hoisted to the surface. The rubbish accumulated in the space from which the vein has been removed serves to support the walls and keep them from coming together. The annexed cut (Fig. 16) shows, in a transverse section, the method of supporting the rubbish and the floor of the level beneath. The workman, while engaged at stoping, stands on a platform laid on stulls, as indicated in Fig. 15 above. Stoping upwards is the most rapid and economical method, when the ores are not very valuable, or difficult to distinguish from the rubbish. Of course, in working from below upwards, the looseness and working away of the rock is aided by gravity, the fragments tending downwards by their own weight, and it is therefore easier for the miner; but, on the other hand, as the rock and ore are thrown together by each blast upon the pile of rubbish below, it requires care, on the part of the miner, to select out the valuable portion of the lode without sending up too much of the worthless rock, which, apart from the cost of hoisting it to the surface, is needed to fill up the space left by the previous excavations, in order to help to support the walls.

Fig. 16.

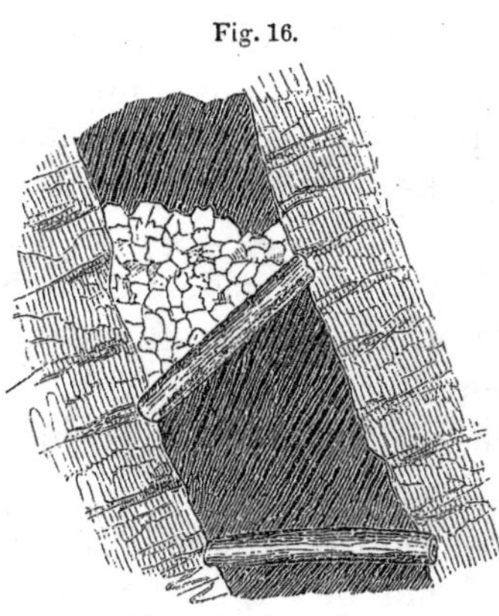

Timbering in a mine.

width, the process of removing their contents becomes a more difficult one, especially if the surrounding rocks are not firm and unyielding. The general method, however, is the same: the work must be subterranean and not open to the day. One or more shafts must be sunk into or near the mass to be removed, and the work commenced at as low a point as possible. Levels are then driven into the ore, and as fast as the valuable mineral is removed, the rubbish which is made is piled up in its place, so as to support the portion left overhead. If there is not enough material for this, stones must be sent down from the surface. In this way the whole contents of a mass may be removed without loss. The methods pursued in such workings must be specially adapted to the shape of the deposit of ore, the degree of solidity of it and of the adjoining rock, and a variety of other circumstances; the general principles, however, are those above indicated.

CHAPTER II.

GOLD, PLATINA, AND SILVER (IN PART).

SECTION I.

MINERALOGICAL OCCURRENCE AND GEOLOGICAL POSITION OF GOLD.

MINERALOGICAL OCCURRENCE.—Gold occurs, in nature, in the following forms :—

Alloys.

Native Gold.—An alloy of gold and silver, with traces of iron, copper, and other metals.

Porpezite.—Gold and palladium. (*Ouro poudre.*)

Rhodium Gold.—Gold and rhodium.

Amalgam.

Gold Amalgam.—Gold and mercury, with a little silver.

Ores.

Graphic Tellurium.—Telluride of silver and gold.

Aurotellurite.—Telluride and antimonide of gold, silver, and lead.

Of the above-named substances, the first, native gold, is that form of combination in which almost the whole amount of this metal obtained in the world is found. Of the others, graphic tellurium is the only one which exists in sufficient quantity to be of economical importance, and that only in one district; the others are exceedingly rare substances.

Native gold is invariably found alloyed with silver; no analysis has ever been made of it which has not given a small amount of this metal: besides the silver, there are traces of other metals, among which iron and copper are rarely wanting, although generally present only in minute quantities. The following table will exhibit some of the analyses of native gold from different localities.

Locality.	Analyst.	Gold.	Silver.	Copper.	Iron.	Sp. Gr.
RUSSIAN EMPIRE.						
Schabrowskoi washings,	G. Rose,	98 96	·16	·35	·05	19 099
Boruschkoi washings,	G. Rose,	94·41	5·23	·36 with iron and loss		18·440
Beresowsk mine, in brown hematite,	G. Rose,	93·78	5·94	·08	·04	
Beresowsk, in quartz, .	G. Rose,	91·88	8 03	·09	trace	
Zarewo-Nicolajewsk, near Miask, washings,	G. Rose,	89·35	10·65	trace	trace	17 484
Alexander-Andrejewsk, near Miask, washings,	G. Rose,	87·40	12·07	·09	trace	17·402
Petropawlowsk washings,	G. Rose,	86·81	13·19	trace	trace	16·869
Boruschkoi washings, .	G. Rose,	83·85	16 15	trace	trace	17·061
TRANSYLVANIA.						
Vöröspatak, in porphyry with quartz,	G. Rose,	60·49	38·74	·77 with iron		
Sta. Barbara mine, at Füses, scales in porphyry with quartz,	G. Rose,	84·89	14·68	·04	·13	
AFRICA.						
Senegal, ...	D'Arcet.	86·97	10·53			
AUSTRALIA.						
Bathurst, ...	J. H. Henry,	95·69	3 92		·16	
Locality not given, . .	A. B. Northcote,	99·283	·437	·069	·203*	
South Australia, . . .	A. S. Thomas,	87·78	6·07		6·15	
SOUTH AMERICA.						
Antioquia, New Granada, washings,	Boussingault,	64·93	35·07			14·149
Brazil, ...	D'Arcet,	94·00	5·85			
Marmato, ...	Boussingault,	73·45	26·48			12·666
CANADA AND UNITED STATES.						
Rivière du Loup, Canada,	T. S. Hunt,	86·40	13·60	trace	trace	15·761
Georgia, ...	W. W. Mather,	95·579	4·421	trace	trace	
North Carolina, Lewis mine,	F. A. Genth,	65·03	34·18			14·90
American Fork, scales, .	Rivot,	90·90	8·70		0·20	15·70
Feather River, scales, .	Rivot,	89·10	10·50		0·20	17·55
Locality not given, . .	Henry,	90·01	9·01	·86		15 96

* Bismuth, ·008.

As the undisturbed and unaltered Silurian rocks exhibit, at most, only traces of gold, it appears that this metal only becomes evident after the strata have been metamorphosed, or invaded by igneous and eruptive rocks. The period at which the segregation of the auriferous veins took place, and that of their impregnation with gold, still remain open questions, even after the age of the formation in which they are included has been determined. In general, the quartz veins may be presumed to have originated at the time of the metamorphic action on the strata themselves; and where there are igneous rocks in the immediate vicinity, the development of the metallic contents of the adjacent veins is usually ascribed to their presence, since it is so often found that the metalliferous deposits are intimately associated with eruptive masses. It is not necessary to suppose, however, that these phenomena throughout the world were confined to any particular geological epoch; on the contrary, there may have been a repetition of similar conditions at periods of time very distant from each other.

Murchison has shown that the impregnation of the rocks of the Ural with gold took place at a very recent geological period, as late even as the drift-epoch: this is one of the most striking facts developed by this distinguished geologist in his great work on Russia. In other important gold regions we have not sufficient data for fixing with much definiteness the time of the concentration of the metal into veins. In the Southern United States, it appears that no very great change in the character of the strata has taken place since the epoch of the new red sandstone; and the date of the gold-bearing veins may with probability be assigned to a period between that formation and the carboniferous. In Australia, the elevation of the auriferous strata and their probable impregnation with gold seem to have been later than the epoch of the coal. In regard to the age of the quartz veins of California, but little positive information can be given. Analogy and all the facts thus far obtained, lead us to suppose that the auriferous masses are included in slates of palæozoic age, highly metamorphosed; but of the period of the igneous action which may be supposed to have been the cause of their impregnation with gold, we know but little. In the Cordilleras of South America, the period of metalliferous emanations seems to have been after the deposition of the cretaceous strata; and disturbances of the rocks, which may have been attended with phenomena of this character, have evidently continued down to a very recent geological period, as well here as in California.

It is not, however, from workings in the solid rock that the principal portion of the gold of commerce is derived. Probably nine-tenths of it, at least, are obtained from *gold-washings*, or the separation of the metal from the superficial detritus which lies upon the rock in place, and which is included by geologists among the drift and alluvial deposits. Nature has performed the washing and concentrating processes herself on a large scale, and accumulated the precious metal in a position from which it can be obtained with facility, without any considerable outlay of skill or capital. It is this circumstance which causes such extraordinary fluctuations in the production of gold, since an almost unlimited quantity of unskilled labor can be at once applied to its collection, while the same amount of metal, were it remaining in its original position, could only be acquired by the application of a vast amount of capital. Indeed, the experience of the past has shown very clearly, that had not nature effected this concentration, the larger portion of the auriferous veins could not be worked, since the metal is too much scattered through them to be separated with profit.

The separation of the gold from its original matrix and its deposition among the strata of gravel, sand, and clay, or beneath them upon the surface of the rock, has been the result of causes acting through an immense period of time; and which have not yet ceased to operate, although their energy seems no longer equal to what it must have been at a former epoch. The rocky strata of the earth are constantly undergoing abrasion from the combined action of various meteorological causes; of which one of the most powerful at present is the alternate freezing and thawing of water in fissures and cavities, which tends to wear away and disintegrate the most elevated portions, especially of the slaty beds, and to carry down the abraded and loosened materials, and spread them out in the adjacent valleys. In lofty and rugged mountain chains, where torrents of rain frequently fall, and the streams, suddenly swollen to a great volume, rush with tremendous violence down rapidly declining valleys, their force becomes capable of wearing away the rocks with great rapidity. This mechanical action is frequently aided by a chemical one; the strata undergoing a molecular change which softens them and renders their abrasion easy. As the enclosing rocks are thus worn away, the quartz-veins become disaggregated by the oxidation of the iron they contain, and are themselves crushed into fragments and borne down into the valleys, where the metallic particles, having by far the highest specific gravity, are first deposited and sink to the bottom, while the lighter earthy portions are carried farther.

The exact identity in origin of the auriferous gravel and sand of various gold regions and the modern alluvial formations is not admitted by all geologists. Murchison, especially, insists strongly that the superficial deposits which contain gold are in no way to be confounded with detritus formed by present atmospheric action, but rather that they were the result of diluvial currents connected with and originating in physical changes in the surface of the globe, such as the elevation of mountain chains. A source as vast as this seems to be required to account for the accumulated masses of loose material which are scattered over the Siberian plains, or at the base of the Cordilleras; and it seems reasonable to admit that the elevation of the Ural and the Andes may have been connected with the deposition of the auriferous deposits which lie upon their flanks. But until geologists shall have more satisfactorily settled the relations between the drift and alluvium, and the transition from one to the other, and until a much more thorough exploration shall have been made of some of the greatest gold-producing districts, it will be a hazardous matter to enter into any discussion of the subject. It is most probable that the causes by which the auriferous detritus was accumulated must have been similar in quality, if not in quantity, to those now in action; and when we take into account the immense period through which they must have been at work, perhaps a less degree of intensity will be thought necessary to account for their effects.

It will be indeed an interesting fact, if it should be proved that the largest amount of gold is to be obtained in regions where the rocks have been most recently and extensively invaded and uplifted by igneous masses; and it might furnish a clue, among other things, to the fact of the vastly greater accumulation of gold in the superficial débris of California than in those of the Southern Atlantic States, although the richness of the auriferous veins in place does not seem to be much, if any, greater in one region than in the other.

SECTION II.

GENERAL DESCRIPTION OF FOREIGN GOLD REGIONS.

HAVING prefaced by some general remarks on the occurrence of gold, we proceed to illustrate the subject more fully before entering upon the gold-regions of the United States, by giving a condensed account of some of the principal auriferous districts of other parts of the world. The geographical order which will be followed in this, as in the succeeding chapters, will be: to commence with Northern Europe, including the northern part of Asia embraced in the Russian Empire; Scandinavia; England; Central Europe; France; Spain; Southern Europe; Northern Africa; Central Africa; Central Asia; Southern Asia, and the East India Islands; Australia; South America; Central America, and Mexico; Canada, and the United States from the northeast towards the southwest.

GOLD REGION OF THE URAL AND SIBERIA.—The gold of the Russian Empire is obtained almost entirely from the eastern slope of the Ural Mountains, from Siberia, and in the Caucasus.* In Russia in Europe a very small quantity only is obtained, from the western slope of the Ural. A small amount of this metal was formerly obtained in the government of Archangel, but this locality has been abandoned

* Consult TSCHEWKIN in Russian Journal des Mines; and G. Rose, "Reise nach dem Ural."

since the commencement of the present century. In Asiatic Russia, the auriferous districts are principally included in the governments of Perm, Oremburg, Tomsk, Yenisseisk, Irkoutsk, and the district of the Kirghese. The first discovery of gold was made near Ekatherinenburg, in 1743; the first workings were in the solid rock, and were commenced in 1752, at the mines of Beresow, which are still productive, though in a very diminished degree; their yield, in 1850, being less than 100 pounds. In 1823, there were sixty-six localities in the Ural where gold had been mined from the solid rock; but they had, with the exception of eight, all been abandoned. The veins are numerous at the Beresow mines; they are contained in granite, which itself forms veins in the talcose, chloritic, and micaceous slates.* The productive ones are of quartz, and they cut the granitic masses at right angles, having a nearly vertical dip. They do not generally extend beyond the limits of the granite. The working seems to have demonstrated that the amount of metal decreases in depth. During the latter part of the last century, the Beresow mines yielded from 600 to 800 pounds of gold annually; and the stamped ore yielded from five to eight zolotniks of fine gold to the hundred poods; equal to ·0013 to ·00208 per cent.

The proper gold washings of the Ural, which have produced such large amounts of this metal, and which had acquired such a high celebrity before they were eclipsed by those of California and Australia, were commenced by the crown in 1814. Those of Western Siberia were established in 1829; of Eastern Siberia in 1838. The following table shows the produce of these washings from the commencement.†

	lbs. troy.
1814 to 1820, product of the crown washings, . .	1,085
1820 to 1830, crown and private, " . . .	73,200
1830 to 1840, " " " " . . .	175,460
1840 to 1850, " " " " . . .	553,955
And of gold obtained from mining in the rock, from 1752 to 1850, ...	128,570
Total, . . .	932,270

* G. Rose, "Reise nach dem Ural," i. 175.

† Tschewkin, Jour. des Mines, quoted in Ann. des Mines, (5) iii. 805.

Since 1847, when the produce of the Russian mines seems to have reached its maximum, there has been a decided diminution; few new localities have been discovered, and in the Siberian washings the yield of the sand is lessening, and also the amount of gold obtained; in the Ural, on the other hand, owing to the perfection of the apparatus employed, and the skill with which the works are conducted, the produce of gold is increasing, although the tenor of the sands has diminished to ¼ zol. per hundred poods, equal to ·00006 per cent. It may be safely inferred that the Russian produce of gold has reached its maximum, and that it will continue slowly to decline.

The washing machinery of the Ural mines is remarkably efficacious, much time and money having been expended on its perfection; were it not such it would be impossible to wash the auriferous sands with profit, since their tenor in gold is very low, the average not being more than from 4 to 6 parts in a million, equal to about 3½ grains per bushel of 100 lbs. av.

Great Britain.—The number of localities in England where gold has been found is very considerable, but there is no reason to suppose that it occurs in large quantities. In South Wales the Romans mined extensively in the Silurian rocks, with but little return of metal. In Dumfriesshire, in the Lead Hills, several hundred men were employed in washing gold sands during the reign of James V. The occurrence of this metal in the tin stream-works of Cornwall and Devon has been observed for hundreds of years, and a few ounces are still obtained yearly by the miners,* which are mostly preserved as a mineralogical curiosity. Borlase mentions a piece weighing 15 dwts. 3 grs.

The Wicklow Mountains, in Ireland, are made up of rocks which might, from their geological nature, be presumed to contain gold; and, in fact, this metal has been obtained there in small quantity; one piece weighed twenty-two ounces, and quite an excitement was raised on the subject.†

* De la Beche, "Survey of Cornwall," p. 614.

† E. Hopkins, in Mg. Almanac for 1849, p. 195.

having their worn-out parts exchanged for others, without much delay or cost. In the construction of the latest invented "gold-quartz crushing machines," it seems as if this circumstance had not been sufficiently attended to.

If these machines are capable of separating the gold with as much economy as they profess, there can be no doubt that there are numerous mines in England which furnish some ore which will pay for working; but, in general, the gossan, which is supposed to be the richest gold ore, does not continue to a very great depth, usually not more than twenty fathoms; and unless the lodes are very wide, the quantity which they would furnish would be small in comparison with that which would require to be stamped in order to furnish a large supply of gold.

Mr. John Taylor, Jr., whose experience in these matters is very great, remarks on this point* as follows: "I have seen evidence to make me believe that British gold ores, in moderate quantities, can be obtained, and that, if they are skilfully treated, they can be made to yield a moderate profit; but, beyond this, I cannot persuade myself that producing gold in England will ever be a large or very lucrative branch of industry."

GERMANY.—*Austrian Empire.*—The amount of gold obtained in Germany is very small indeed, although in some places mining or washing has been pursued almost uninterruptedly since the days of the Romans.

Hungary, Transylvania.—The exploitation of the mines of gold, silver, lead, and copper, of Hungary has been carried on almost uninterruptedly since the eighth century; and the results of these long-continued labors present one of the most interesting fields of investigation for the mining engineer which can be found in the world. At Schemnitz, Kremnitz, Neusohl, and Libethen, in Lower Hungary, the visitor may study the results of centuries of experience in the slow development of the processes for the treatment of auriferous and argentiferous ores.

There are mines of gold and silver, as well as copper and

* Eng. Mining Journal, Dec. 1853.

The different provinces of the Austrian Empire furnished each the proportional amount of gold specified in the following table, according to the average of the official returns of the years 1840 to 1847:—

Transylvania, ...	53·3 per cent.
Hungary, ...	45·6 "
Salzburg, ...	0·6 "
Tyrol, ...	0·25 "
Styria, Bohemia, &c., . . .	0·25 "
	100·00

The total production of the empire amounted, in 1848, the last year for which returns have been published, to 5645 lbs., having steadily, although slowly, increased from 2682 lbs. in 1820. Its present yield may be estimated at 6000 lbs.

FRANCE.—*The Rhine.*—The sands of the Rhine are still washed, on a small scale; formerly, there is good reason to believe, the production was very considerable. Historical documents show that, in 667, the right of washing the sands of the bed of the Rhine was considered to be of value. The present yield (1846) is estimated by Daubrée at 45,000 francs. The auriferous portion of the bed of the river is included between Basle and Mannheim, but the washings have been especially numerous in the vicinity of Strasburg.* The sands recently washed yield from thirteen to fifteen hundred-millionths; they rarely contain more than seven ten-millionths. The washer makes, on an average, from one and a half to two francs a day, and sometimes as much as ten or fifteen francs. Daubrée considers the average yield of the sands of the Rhine, Siberia, and Chili, to be in the proportion of 1 : 10 : 37.

There are several localities in France where gold has been produced in small quantity. Diodorus speaks of the richness of the Gallic rivers in golden sands, but nothing at present existing would lead to the opinion that the amount obtained was ever very great. The river Ariège derived its name from its yield of auriferous sands (*aurigera*), and up to

* Daubrée, Comptes Rendus, xxii. 640.

there are reports of the recent discovery of rich gold sands. The washings at present are principally confined to the rivers Sil and Salor, of which the whole produce may perhaps amount in value to $8,000 per annum.

ITALY.—Quite a number of localities were known to the ancients as producing gold. The Po is still one of the gold-bearing streams. The only works of any consequence at all are in Piedmont and Savoy. In the Anzasca Valley, near Monte Rosa, an auriferous pyrites was worked until quite recently, which yielded only from 2 to 85 francs per cwt. of ore, or 8 dwts. of gold to the ton. The amalgamation works were scattered up and down the valleys on the small streams. The whole amount produced in the province of Ossola, to which these works belong, was, in 1829, about 250 lbs. troy, with a profit of not much over $15,000. These mines were extensively worked in the time of Pliny; and, according to him,[1] the Senate fixed 5,000 as the number of slaves who were to be allowed to work in them, lest the price of the precious metal should be reduced.

A French mining company was also recently engaged in working a gold mine in serpentine, near Genoa, on the flanks of the Col Bochetta.

CENTRAL ASIA.—Golden sands had frequently been discovered in the Caucasus; and, in 1851, the Russian government took steps to establish washings there.* In that year the deposits of auriferous sands occurring on the branches of the Koura, which originate in the great trans-Caucasian chain, were examined, and the region was found to be in all respects identical in character with the gold-bearing districts of Siberia. Evidence was obtained that these sands had been worked previous to the Christian era. On the River Akstafa, the auriferous bed consists of rolled pebbles and boulders, mixed with quartz, brown iron ore, and clay, the whole covered with six feet of unproductive alluvium.

Thibet is generally supposed to be rich in gold, as well as other metals. The rivers of the western portion of that country are especially referred to as abounding in auriferous

* Ann. des Mines, (5) iii. 830.

sands. Jacobs, who gives the only estimate of the amount produced which I have been able to obtain, fixes it at 10,000 ounces.

Southern Asia.—There can be little doubt that the portion of the Asiatic continent south of the great Himalayan chain has, in former times, yielded vast amounts of gold. The much vexed question of the locality of Solomon's Ophir has not been definitely settled; but it seems not improbable that it was on the Malayan peninsula or some of the adjacent islands. Ritter places it in Hindostan; by others it is supposed to have been somewhere in Thibet. Gold is still obtained, in small quantities, at many localities in India; but it appears that they have long since lost the productiveness which history leads us to believe they must once have had. What metal is now obtained is mostly used for personal ornaments. The washings of the Burrampooter are estimated to yield from 30,000 to 40,000 ounces yearly, by Jacobs and others.*

In the Burman Empire, many streams have gold-washings upon them, which are irregularly worked by the natives, who are represented as especially fond of decorating their persons with golden ornaments; and, their own produce not being sufficient, they are said to import an additional quantity from China for that purpose. According to Jacobs, no reliable data are to be obtained in regard to the amount produced; Mr. Birkmyre, however, puts that of Ava at about 2000 pounds.

Authentic information represents the Malayan peninsula as having been rich in gold, but the amount produced at the present time is small. The character of the inhabitants forbids any extended researches on the part of foreigners, although such have been attempted recently with some success.†

China and Japan.—Of the mining statistics of these countries we have little or no exact knowledge. We know, however, even after making allowance for exaggerations of travellers,

* Jacobs, Historical Inquiry into the Production and Consumption of the Precious Metals, ii. 330.

† Ann. des Mines (5), iii. 516.

were christened "Ophir," and soon after, the Turon Valley was found to be auriferous for a distance of one hundred and thirty miles; twenty miles north of the Turon, on the Meroo, a native shepherd discovered blocks of quartz, rich in gold, lying on the surface, from one of which 60 lbs. of the metal were taken. The next great discovery was at Araluen, two hundred miles south of the Turon; and between these localities numerous others have since been found, extending over a distance of seven hundred miles in length, north and south.

The name of Victoria was given, in 1850, to a district previously almost unknown, which lies around Port Philip, and of which Melbourne was the principal settlement. In this colony the discovery of gold was announced in August, 1851, and the Ballarat diggings, about fifty-five miles north-west of Geelong, at the junction of the slates and the trappean rocks, were soon found to surpass all that had been discovered in New South Wales. This, and the Mount Alexander gold-field, a few miles north, have yielded almost fabulous quantities. In October, 1851, about 3,000 persons were at work at the Ballarat diggings; but, in December, it was computed that at least 12,000 persons were collected at the Mount Alexander gold-field, within a space of fifteen square miles. It is not necessary to give particulars of the large nuggets found, or the excitement which followed on each successive discovery; the statistics of the yield of these gold-fields are sufficient to enable every one to judge of the effect which such revelations of golden wealth must have had on the astonished Australians.

The Australian gold is of remarkable fineness, the different analyses giving from three to seven per cent. of silver. The whole amount, almost without exception, has been thus far obtained from washings; and although numerous quartz-mining companies have been organized, they have, it is believed, without exception, failed to accomplish anything. Mr. Calvert, in his work, which draws so heavily on the credulity of the reader, declares. that he had discovered nearly two hundred and thirty-eight gold-bearing quartz veins, producing on the average 2½ dwts. to the ton. He speaks of one, in particular, the Macquairie vein, as having been traced forty miles, running north and south. This vein, he admits, could not probably be worked with profit at the present time. From another vein, Mr. Calvert mentions that he broke off 3 cwts. of rock, which yielded 76 lbs. of the pure metal. The investigations now going on in London, with regard to the gold-quartz mining companies of Australia, would indicate that they are mostly swindling concerns; and also that the quartz veins are not rich in gold, or, at least, have not yet been proved to be.

The auriferous deposits present the most striking analogy with those of California. There are innumerable "diggings," which are sometimes quite superficial, and sometimes extend through the detritus and loose materials to the "bottom rock" or "ledge," namely the rock in place. In the surface-workings, which are exclusively on the flanks of the hills, the gold is diffused through the gravelly soil to the depth of six to twelve inches, beneath which there is a stiff, red clay, which contains little or no gold. In the deeper workings, it is necessary sometimes to sink twenty-five or thirty feet, or even more, before reaching the auriferous deposits. These are of varying character. The channels of the small streams coming down from the mountains, and often entirely dry for most of the year, are usually rich in gold, which is found accumulated against the "bars" or projecting ledges of rock, and forced into the chinks between the strata. The more precipitous the torrent, the larger the nuggets which are found; where the valley expands out, the golden particles are smaller. These rich deposits are frequently found at the entrance of a lateral into the main valley, where there has been an eddy in the current, caused by the contraction of the mouth of the lateral valley.

In the valleys of the Bendoc and Delegete, according to Rev. W. B. Clarke, the superficial deposits are in the following order:—

1. Gold-bearing detritus, made up of fragments of slate and quartz, cemented by argillaceous matter.
2. Pipe-clay.
3. Erratic blocks and pebbles of quartz containing gold.
4. Rock in place.

The minerals and rocks associated with the gold deposits are similar to those of other great gold-bearing districts. Quartz is exceedingly abundant, and evidently the gangue of the gold. Magnetic iron sand, sometimes titaniferous, is rarely wanting. The rarer minerals are topaz, garnet, zircon, corundum, rutile, and the diamond, of which one crystal, at least, is said to have been found on the Turon. Platina is not wanting; but, thus far, seems to be quite rare. The quartz and oxide of iron were evidently originally associated with the gold; the precious gems may or may not have occurred in the same connection; their hardness, indestructibility, and high specific gravity, would cause them to be found under the same circumstances as the gold, and in company with it, in the superficial detritus, even if derived from rocks at a considerable distance from the gold-bearing quartz veins.

The question, whether the accumulations of superficial detrital matter containing gold, such as pebbles, fine clays, and sands, are due to the long-continued and gradual action of causes which are still in operation, such as currents of water, rain, and the like, or whether they are the result of cataclysmal action, namely, of a drift-period similar to that of our "Northern drift," is one of great interest both scientifically and practically; but the data are too incomplete, at present, to allow any positive opinion to be formed on that point.

Mr. Stutchbury, one of the government geologists in New South Wales, considers that the evidence is decidedly in favor of the gold-bearing detritus having been accumulated by the slow disintegration and consequent denudation of the rocks, under the action of causes not differing from those now in operation.

The apparatus used by the miners for separating the gold is of the greatest simplicity, consisting of the cradle and pan. In the richest fields of Victoria, up to a very recent period certainly, thousands of miners have collected large quantities of gold without even so simple a machine as the cradle, by simple "panning."

Imperfect as are the statistics of the total yield of Australia, they are far more satisfactory than the accounts of the average contents of the auriferous earth washed. Mr. Calvert estimates it at $\frac{9}{10}$ grain in a cubic foot, which he says will seem very low. Considering the average depth of the auriferous beds to be thirty-nine inches, and the area covered by them to be equal to 68,700 square miles, the total amount of gold would be 434,191 tons, worth, at £3 19*s.* per ounce, the trifling sum of £46,100,571,660. To this Mr. Calvert adds, that this amount is exclusive of the gold contained in the solid rock and the quartz veins.

The difficulty of arriving at an accurate knowledge of the amount of gold produced in Australia seems to be very great. The following estimates are selected from those supposed to be most reliable.

The gold produced in the colony of Victoria is mostly brought, under government or private escort, to Melbourne, Geelong, or Adelaide; but a portion of it is carried by private hand across to Sydney. That of New South Wales, which is much less in quantity, goes to Sydney almost exclusively. Hence, we have a first approximation to the total produce in the amount carried under escort to these sea-ports. But this is but a portion of the gold, since a considerable quantity finds its way to the cities through private hands. The amount exported and manifested is a better guide; but to this must be added what is taken by passengers privately, and what remains in the country, in circulation, in the hands of bankers, and in transitu.

The returns of the exports of native gold for the different years are believed to be approximately as follows:—

Year.	Victoria.	New South Wales.	Total of Australia.
	lbs. troy.	lbs. troy.	lbs. troy.
1851, ...	12,176	11,910	24,086
1852, ...	164,580	70,636	235,216
1853, ...	212,105	63,800	275,905
Total since the discovery of the gold to end of 1853, .	388,861	146,346	535,207

The production of 1852 is that which can be most accurately determined; and here there are numerous estimates which agree tolerably with each other.

The total production of the colony of Victoria is estimated as follows by different authorities:—

	1851.	1852.	1853 (1st half).
English Mining Journal, .	224,140 oz.	4,167,571 oz.	1,296,059 oz.
Mr. Khull, bullion-broker at Melbourne, ...		4,247,657	
William Westgarth, . . .		4,545,780	
Melbourne Argus, ...	4,000,000 (1851 and 1852)		

The total yield of Victoria and New South Wales for 1852 is estimated by M. Delesse, from data obtained by Murchison from official sources, at 400,000,000 francs = £16,000,000.

Mr. Birkmyre, who was on the spot, and may be supposed to have had good opportunities for judging, has placed his estimates somewhat lower, namely:—

1851, ...	23,667 lbs.
1852, ...	266,178 "

These estimates, however, seem rather too low, since the quantity actually exported in 1851 seems to have exceeded that given by Mr. Birkmyre.

During the year 1853 the produce diminished considerably. The yield of New South Wales reached its maximum early in 1852, apparently. That of Victoria seems to have been at its highest point from July to November, 1852, when Mount Alexander and Ballarat were producing enormously, as shown by the following account of the gold taken from those fields to Melbourne and other places, according to escort returns:—

June,	1852, ...	162,990 ounces.
July,	"	353,182 "
August,	"	350,968 "
September,	"	366,193 "
October,	"	264,683 "

The following figures show the amounts taken from the Victoria gold-fields for each month of the years 1852 and 1853, specifying for the first nine months of each year the number of ounces brought by escort to Melbourne, and for the last three months, to Melbourne and Geelong, thus affording a comparative view of the yield of the two years:—

		1852.	1853.	
Brought to Melbourne, ...	January, .	53,594	186,615	ounces.
" "	February, .	56,142	172,329	"
" "	March, . .	62,026	169,654	"
" "	April, . .	68,041	170,427	"
" "	May, . .	77,247	116,812	"
" "	June, . .	116,009	122,695	"
" "	July, . .	320,218	198,007	"
Brought to Melbourne and Geelong,	October, .	334,648	200,280	"
" " " "	November,	327,623	162,393	"
" " " "	December,	121,827	176,580	"

The same inference may be drawn from the following table, which is stated to have been prepared by a gentleman in South Australia, who had given much attention to the statistics of gold.* It shows the amount of gold brought to Melbourne, Geelong, and Adelaide, by government and private expresses, and also gives an estimate of the quantity brought by private hand, together with the shipments for each quarter from the first discovery to the middle of 1853:—

Year.	Quarter ending	Brought by escort.	Estimated by private hand.	Total.	Shipped.
1851, . .	Dec. 31, ...	124,522	99,618	224,140	145,116
1852, . .	March 31, . . .	178,006	356,012	534,018	420,445
" . .	June 30, ...	299,615	599,230	898,845	339.729
" . .	Sept. 30, ...	1,017,283	508,641	1,525,924	512,692
" . .	Dec. 31, ...	967,027	241,757	1,208,784	702,110
1853, . .	March 31, . . .	576,192	144,048	720,240	661,957
" . .	July 2, ...	460,655	115,164	575,819	572,121
		3,623,300	2,064,470	5,687,770	3,354,170

The principal yield of Victoria, up to the middle of 1853, was from the Ballarat and Mount Alexander gold-fields; several others have been discovered, but none of them have compared in richness with these two. As might reasonably have been expected, the large number of persons congregated there, and employed in digging and washing, amounting to over 100,000 probably, caused these limited areas of unpre-

* Eng. Mg. Journal, Nov. 5, 1853.

cedented richness to be soon worked out. New localities may be discovered, and the old ones will continue to be worked with more moderate returns; but it seems probable that there will be a gradual falling off in the yield. This is the opinion of Mr. E. Hopkins, an English mining engineer well known to the scientific world, who also gives it as the result of his observations in Australia, that there are no quartz veins there worthy of being worked.

The amount of gold imported into England from Australia is stated as follows:—

1852,	£7,282,635
1853,	14,972,743

Of course the amount received in England in any particular year, would not be a guide to that actually obtained during the same year in Australia, the length of time required for the passage being several months.

On the whole, after a careful consideration of the various returns and estimates, some of which evidently bear the marks of exaggeration, while others appear to have been carefully compiled, I have adopted as the production of Victoria and New South Wales together, the following amounts of pure gold, which probably nearly approximate to the truth.

Year.	lbs. troy.
1851,	30,000
1852,	330,000
1853,	210,000

South America.—Although the silver mines of South America were at one time pouring forth a stream of wealth unparalleled in the history of the world, and causing a general rise in prices, similar to that which we are now witnessing, as the result of an unprecedented yield of gold; yet, of this latter metal, the quantity furnished by that country has never been very large. In 1800, the whole produce of the Southern American continent was estimated by Humboldt at 33,524 lbs. of gold; of which 9,900 lbs. were the yield of Brazil; while at the same time the annual proceeds of the silver mines, according to the same authority, amounted to 691,625 lbs., gold and silver being as 1 to 21 nearly by weight. In 1850, the amount of gold had fallen off to about 24,000 lbs., or three-fourths of what it was at the commencement of the century. According to Chevalier, the total yield of the precious metals from the Andes of Peru and Bolivia was, up to 1846, in the ratio of 1 part of gold by weight, to 170 of silver.

New Grenada.—The ancient vice-royalty of New Grenada was, from 1819 to 1831, united with Venezuela, and part of the time with Equador, forming the republic of Columbia. Each of these states is now nominally an independent republic. This part of South America, as also Brazil, produces almost exclusively gold. Of the New Grenada mines we have ample accounts, although not of recent date, from Boussingault, Chevalier, and other travellers and mining engineers, sent out to explore that region; the principal mines and washings are in the provinces of Antioquia and Veraguas. In the former there is said to be hardly a river which does not flow over auriferous sands, and there are also numerous quartz-veins in the granite which are worked to some extent. The principal mines of this character, worked in 1850, were on the river Porce. The veins are of quartz, carrying auriferous pyrites, which, near the surface, is much decomposed. They resemble in every respect the gold-bearing quartz-veins of other parts of the world. In the provinces of Panama and Veraguas the veins are also numerous; but their yield of gold is small, not amounting, according to Boucard,* to over 0·002 per cent., on the average.

The yield of the province of Antioquia, in the year 1847–8, was supposed to be 12,500 lbs. Chevalier estimated the amount annually produced in the republic at 13,276 lbs., and the total produce as follows:—

Previous to 1810, . . .	$295,000,000
From 1810 to 1846, . . .	81,500,000
	$376,500,000

Equal to 556,840 kilos. pure gold, or 1,492,331 lbs. troy.

Mr. Danson,† from information based on the returns of

* Ann. des Mines (4), xvi. 377. † Jour. Stat. Soc. of London, xiv. 40.

the English consuls, estimates the total amount produced, from 1804 to 1848, at $204,085,328.

There is abundant evidence that New Grenada is still rich in gold; under a different climate, and with more energetic inhabitants, its produce of this metal would be greatly increased.

There are several English companies working gold mines in this country.

West Grenada, or Veraguas Mining Company.—This property has been very favorably reported on by a gentleman formerly connected with the St. John del Rey mines. The gold-bearing quartz is said to be very rich. Preparations were making, in 1853, to work it on an extensive scale.

Mariquita and New Grenada.—This company owns the Marmato mines, among others, and is working them with considerable profit.

The New Grenada Mining Company, in the district of Antioquia, has recently purchased the mines of Frontino, nine leagues west of the city of Antioquia. The produce of this mine, as worked for several years by the inhabitants, has been from 15 to 25 lbs. of gold per month. The workings have been confined to one quartz vein, which is mixed with iron pyrites, and varies in thickness from two inches to five feet.

Venezuela.—That part of the ancient viceroyalty of New Grenada which is now Venezuela had produced very little or no gold up to very recent times;* but lately this metal has been obtained in considerable quantity in the canton of Upata, province of Guyana, and also in the province of Cumana, at Campano.†

Brazil.—Although Mexico and Peru have furnished the larger part of the metallic treasures of the New World, their yield has been chiefly of silver. Brazil, on the other hand, has, for a great length of time, been famous for its gold alluvia, which have been worked since the beginning of the last century, and have produced an enormous amount. There is considerable and more precise information than is usually to be had in regard to South American mines, in the works of Von Eschwege, formerly Director-General of the Brazilian mines, Burat, and others.

It was chiefly from the washings of the auriferous alluvia of Minas Geraes, that the large quantity of gold which

* Ann. des Mines (4), xviii. 107.

† Ann. des Mines (5), i. 600.

flowed from Brazil in the eighteenth century was obtained; but, at present, the principal source of produce are the veins in the solid rock. Burat has given an excellent account of the mining region of Brazil,* which shows that the auriferous deposits are somewhat different in character from those in other parts of the world. The rocks are supposed to be of palæozoic age, but are so metamorphosed as to be not referable to any precise epoch. They have not been elevated into Cordilleras, but form a series of disconnected elevations; the trappean and other eruptive rocks appearing rather in dome-like protuberances and dykes than in mountain chains. It seems pretty certain that there have been no geological disturbances of so late a period as the elevation of the Andes. The rocks most developed are gneiss, and those varieties of quartz-rock which are known in Brazil by the names of itacolumite, jacotinga, and itabirite.

These rocks are characteristic of Brazil. The itacolumite is a quartz mixed with chlorite, and is sometimes of enormous thickness. When it takes into its composition specular iron, it becomes the rock known as itabirite or jacotinga, according as it is crystalline or compact. The gold is disseminated through the whole of the metalliferous bed, there being no regular veins; but it is especially concentrated in the vicinity of masses of specular iron, and in connection with quartz.

The great importance of the Brazilian gold mines, as being the only ones both extensively and profitably worked in the solid rock, induces me to present a somewhat detailed description of one of the most important, that of St. John del Rey. This, as well as the other principal companies working in the neighborhood, is owned and managed entirely by English capitalists.

The St. John del Rey Company is divided into 11,000 shares, on each of which £15 have been paid up. The operations were commenced on the estate of Morro Velho in 1834, having been previously carried on in other localities with considerable loss. In 1838 the mine was first worked with profit; and since that period, up to the present time, it has been steadily increasing in value.

The gold is found in a heavy bed of jacotinga, which is intercalated in

* Comptes Rendus, xii. 252.

a very argillaceous itacolumite slate of a reddish-blue color. It had been worked for a hundred years before coming into the possession of the present Company, and was considered exhausted. The auriferous mass averages about 44 feet in width, and dips with the rocks of the vicinity at an angle of about 45° to the southeast. It consists mainly of specular iron mixed with sulphuret of iron, magnetic pyrites, and quartz. Its average yield of gold for the last four years has been as follows:—

1849.	1850.	1851.	1852.	
3·89	4·07	3·89	4·25	oitavas per ton.
·00137	·00143	·00137	·0015	per cent.

In numerous places there are cavities and "vugs," lined with fine crystals of calc. spar and spathic iron, which, themselves, are frequently covered with smaller crystals of specular iron, magnetic pyrites, &c. When these drusy cavities are frequent, the yield of gold diminishes, and this metal is most abundant in the compact, uncrystallized, specular iron. There are three mines known as Bahu, Cachoeira, and Gambu. The principal shafts are sunk inclining about 45° with the auriferous bed, and the ore is raised on a tram-road by cars. The Bahu mine is about 1200 feet deep. As the ore comes up, it is broken by negro women, and then carried to the stamps, of which there are one hundred and twenty heads in operation, moved by three or four large water-wheels. The coarse gold is caught on cow-skins, which are changed every two hours, while the slimes are amalgamated in barrels. In each barrel are placed 80 lbs. of mercury to 16 cubic feet of slime, and the whole is allowed to revolve for thirty hours. The Tyrolian mills have been tried here and found not to succeed.

The Company own 1000 slaves, and employ about 80 or 90 Europeans as overseers, captains, mechanics, and head-men, in every department. Some 300 Brazilians are hired by the day as surface-men. The annexed table shows the quantity of rock stamped, and the amount of gold produced, for a few years past, together with the profits:—

	1846.	1847.	1848.	1849.	1850.	1851.	1852.
Ore stamped (tons), . .	34,935	40,234	58,122	67.336	67,106	79,810	82,642
Gold produced (lbs. troy),	1,465	1,638	2,108	2,473	2,541	3,000	3,233
Net profit, ...	£14,820	21,536	32,269	38,136	35,880	51,586	55,391

The average number of stamp-heads working, during the year 1852-3, was 118·58. The quantity of stamp-sand or slimes amalgamated, was 19,709·59 cubic feet, which yielded 17·02 oitavas of gold (or 2 oz. troy, nearly) to the cubic foot. The loss of mercury was 728 lbs., or 0·037 lb. per cubic foot of slime amalgamated. The net profit in the same year amounted to £55,390 16*s.*, and the total profits of the Company, from 1838 to 1852, to £312,621; a result due mainly to economy and skill in working an ore occurring in abundance, but very poor in gold. The gold contains about 20 per ct. of silver.

Imperial Brazilian Mining Association.—This Company was formed in 1825, for the purpose of working gold mines in the province of Minas Geraes. They commenced by purchasing the estate of Gongo Soco, which in fifteen years produced nearly a million pounds sterling. From 1840, the returns were less satisfactory, the mine for some years not paying its expenses. Recently, however, the accounts are more encouraging, valuable new discoveries having been made, and old workings re-opened.

This Company has 10,000 shares, and £25 per share paid in; it had paid to its stockholders in dividends, up to December 1844, £380,000, having produced in ten years 35,000 lbs. of gold. The Gongo Soco mine was, in 1840, 378 feet deep. The jacotinga at this locality is 50 fathoms in thickness, and softer than that of the Morro Velho mine. The Camara lode, at present worked, is 10 fathoms in width, and composed of a series of quartz layers, or branches, containing iron pyrites, oxide of iron, and oxide of manganese.

National Brazilian Mining Company.—This Company was wound up in 1853, the mine having been ruined by culpable neglect on the part of the manager, who had been superseded. It had produced, in 1850, 120 lbs. of gold, containing 14 per cent. of silver.

The greatest yield of Brazil in gold was about the middle of the eighteenth century. Between 1752 and 1761 the amount on which the royal quint was paid varied from 17,000 to 21,500 lbs. yearly. From that time it gradually fell off, and was in 1822 less than 1000 lbs. The mean annual production from 1810 to 1817 of Minas Geraes, the principal mining district of the country, is given by Humboldt at 4288 lbs. At present the English companies mining in the rock furnish almost all the gold obtained. The washings have nearly ceased. The present annual produce is probably about 6000 lbs.

Chevalier calculates the grand total of gold produced by Brazil, from the earliest period up to 1845, at 3,576,192 lbs. troy. The data on which any such conclusions with regard to the former yield of Brazil are based are of very doubtful character. From 1800 up to 1850 the average seems to have been about $2,000,000 per annum.

MEXICO.—This country is pre-eminently that of silver veins. Gold, however, occurs, hardly in any other way than as a constituent of small percentage of the argentiferous ores. Yet so large is the quantity of silver produced that the accompanying gold becomes a matter of considerable importance. The silver of Guanaxuato and Guadalupe y Calvo is remarkably rich in gold, while that of Tasco, Catorce, and Zacatecas, is poor. The ores are in some instances ground under the arrastras with the addition of mercury, and the silver thus obtained by amalgamation yields from 4 to 6 per cent. of gold. M. Duport calculates that in 1840, or thereabouts, the value of all the gold produced in Mexico, both from the washings and by parting from silver, was equal in weight to about $\frac{1}{135}$ of the latter metal, and, in value, to $\frac{1}{8}$. A considerable amount of gold is also obtained from the washings of Sonora. In the next chapter, when speaking of the silver of Mexico, some statistics of the yield of both the precious metals will be given.

There are gold mines in Oaxaca which have been worked in the solid rock to some extent, and which, in the opinion of Chevalier, are destined to be of importance at some future time.

Central America.—Nothing definite can be ascertained in regard to the gold-washings of Central America. Costa Rica is known to produce a small quantity; but it is mostly smuggled out of the country, so that no estimate can be made of its amount.

SECTION III.

GEOGRAPHICAL DISTRIBUTION OF GOLD IN THE UNITED STATES.

The gold regions of the United States are at least two in number, but of very unequal importance, although similar in extent. The one, that of the Atlantic slope, the "Appalachian gold-field," has been worked to a moderate extent for over thirty years; the other, that of California, "the Sierra Nevada gold-field," has, however, in the six years since it was first discovered, produced more than twelve times as much as has hitherto been obtained on the Atlantic side.

The development of the riches of California reacted upon the gold-bearing districts already known all over the world, causing a general search for the precious metal in many almost abandoned auriferous localities, and stimulating its production in an incredible degree. Among the great dis-

of North Carolina,* on which are noted nine different mining localities, three of which are included by him in the "primary," and the remainder in the "transition or slate."

It was in the summer of 1829 that the first discoveries were made in Georgia; and in 1830 the first gold was sent to the Mint, to the amount, in that year, of $212,000. The first locality known was in Habersham County, and very soon the explorations were extended into Hall and Carroll Counties. The excitement, or "gold-fever," produced by these discoveries was extraordinary. As many as six or seven thousand persons were, soon after, engaged in washing for gold in that region. The researches were carried on, until it was ascertained that the whole of that part of the state lying along the base of the Blue Ridge was more or less auriferous.

South Carolina sent to the mint, in 1829, $3,500; this was the first arrival of gold from that state. Deposits were worked, in 1830, in Chesterfield, Lancaster, and Kershaw Districts. Brewer's mine, in Chesterfield, was one of the most productive localities, there being from 100 to 200 persons employed in 1830 and 1831, who were supposed to average from $1 50 to $3 00 each per day.

The Georgia gold excitement did not last very long; but there, as in the other Southern States where this metal was found, gold-washing continued to be followed by the majority of the washers, not as a regular pursuit, but as one to be taken up and dropped again, as circumstances directed. In 1831, the subject of establishing a Mint in the gold region was agitated, and, as each state claimed one as necessary to the development of its metallic wealth, three were at length built, and commenced operations in 1838; one at Charlotte, N. C.; one at Dahlonega, Ga.; and a third at New Orleans, which latter, however, had, previously to the Californian discoveries, hardly coined any bullion of native production. The production of gold seems to have reached its maximum in the period from 1828 to 1845. From the last-named year it was evidently falling off rapidly, previous to 1853.

* Sill. Am. Jour. Science, xvi. 1.

Professor Tuomey's State Geological Report enables us to form a good idea of the state of the gold mining interest in South Carolina about 1848, the time of its publication. Although a great number of localities were worked, and called mines, they must have been either on a very small scale, or else worked with loss, since the yearly produce of the state, at that time, could not have much exceeded $50,000. The following mines are noticed as the most important ones by M. Tuomey.

Brewer's Mine, Chesterfield District. A bed of quartz of immense thickness, 800 yards in its widest part, is here worked and found to be more or less auriferous throughout its whole extent. The upper part of this bed was so disintegrated, that it was worked for some time under the impression that it formed a part of the superficial deposit of clay, gravel, and pebbles, which was itself rich in gold, and which had been worked from 1838 to 1843. Two hundred persons were employed here in 1848, but the workings were conducted in the most shiftless manner. Other mines were worked on a small scale in this neighborhood.

Catawba and Lynch's Creek Mines, Lancaster, Chesterfield, and Kershaw Districts. The following mines are enumerated here: Lawson's Mine (in North Carolina, near the line), Ezell's Mine, Blackman's, Hale's, Cureton's, Belk's, Perry's, and Stevens's.

Lawson's Mine had been opened over a space of one mile in length, and in some places, to a depth of 40 feet.

Blackman's Mine. A talcose slate furnishes the gold here. The productive portions are enclosed in the barren in lenticular masses. Very extensive workings have been made.

Hale's Mine is of the same character. The mines of this neighborhood had been steadily worked from 1828 up to 1848. They are, however, but little more than a series of open cuts from twenty to forty feet deep.

Fair Forest Mines, Spartanburg and Union Districts. Among these are Nott's, Harman, Fair Forest, West's, and Bogan Mines.

Nott's Mine is opened in an enormous bed of quartz, which is in some places forty feet thick. This is considered by M. Tuomey to be a true vein, and to cross the slates at a small angle. A large quantity of ore has been taken out from an open cut fifteen to twenty feet deep. A perpendicular shaft ninety feet deep was sunk, and connected with the vein by a short cross-cut. Three thousand dwts. of gold were taken from eleven bushels of decomposed ferruginous matter.

Harman's Mine. Copper pyrites is said to form a considerable portion of the quartz bed worked here.

Fair Forest and West's Mines. These are in talcose and micaceous slates, which are decomposed to a depth of from 90 to 100 feet. Below that point

east at a very high angle. In Virginia, they stand nearly vertical.

Canada.—The auriferous district of Canada, according to Logan,* is found to comprehend an area of between 3000 and 4000 square miles. It appears to occupy nearly the whole of that part of the province which lies on the south-east side of the prolongation of the Green Mountains into Canada, and extends to the boundary between the colony and the United States. The gold has been obtained exclusively from washing the superficial deposits, and is not known to have been found in place. The deposit in which it occurs is considered by Mr. Logan to be part of the ancient drift, the "Laurentian" of Desor, alluded to in various reports of the Canada survey as "tertiary" and "post-tertiary," and containing bones and shells of animals of existing species. In the localities where the gold occurs, the coarse materials of the drift are chiefly fragments and rolled pebbles of the clay-slates and gray sandstones on which it rests; but it contains also pebbles and boulders of talcose slate and serpentine, with magnetic, specular, chromic, and titaniferous iron, and masses of white quartz, which are derived from the mountain-range bounding the district on the northwest.

Gold-washings have been carried on, on the Du Loup and Chaudière Rivers, and about 1900 dwts. were obtained during the season of 1851–52, by fifteen men. The largest nugget weighed 2 oz. On the Touffe des Pins, lumps of greater size were found; one weighed as much as 4 oz. The results, thus far, have not been very satisfactory, as far as profitable working is concerned, although it is demonstrated that the precious metal does occur over a very considerable extent of surface.

Vermont.—Gold has been known to exist in Vermont for twenty-five years. I have had specimens of native grains from that State in my possession for a long time. According to Rev. Z. Thompson,† a lump was picked up in Newfane, in 1826, which weighed 8½ ozs. It was pure gold, with exception of some small quartz crystals attached to it,

* Geological Survey of Canada, Report of Progress, 1850–51, p. 6.

† Appendix to Thompson's Vermont, p. 48.

weighing perhaps half an ounce. The specific gravity was 16·5. The gold-formation of this state forms, according to the same authority, a narrow and irregular belt extending through the entire length of the state. The rocks which mark the line of the formation are talcose slate, steatite, and serpentine, accompanied by magnetic, specular, chromic, and titaniferous iron, and also the sulphuret and the hydrous peroxide of the same metal. Rock-crystal is common, and is sometimes traversed by rutile.

At Bridgewater, the gold occurs in quartz associated with the sulphurets of iron, copper, and lead. The quartz is in seams or beds, dipping 55° to the east, and of irregular thickness, not exceeding ten or twenty inches. The quantity obtained here seems to have been very small.

There do not seem to be any well-authenticated accounts of native gold between Vermont and Maryland. A systematic search along the line of the proper formation might reveal its presence, but there is no reason to suppose that it would be found in any considerable quantity. In Pennsylvania, the occurrence of a single grain has been noticed,* but of uncertain locality, and under circumstances which render it doubtful whether it was a genuine specimen oi native gold. According to F. A. Genth, however, the lead and copper ores of Lancaster County contain distinct traces of both gold and platina.

MARYLAND.—Although gold has been found to some extent in Maryland, the amount produced seems to have been, thus far, very trifling. I am not aware of any mines of this metal worked in that state at present. It has been discovered on the farm of Samuel Elliot, in Montgomery County.† The quartz, which forms the gangue, crops out amidst a decomposed talcose slate, so that it is easily mined. The average yield of a portion assayed at the Mint is said to have been at the rate of $522 per ton.

VIRGINIA.—The product of the Virginia gold mines has been small and pretty constant, from 1830 up to the present time, the amount annually deposited at the Mint being

* C. M. Wetherill, Trans. Am. Phil. Soc. N. S. x. 350.

† Proc. Am. Phil. Soc. 1849, p. 85.

between $50,000 and $100,000. Within the last year or two, however, mining has been commenced, on a considerably extended scale, at a number of localities, the result of which will soon determine whether the auriferous quartz is sufficiently rich to be worked with profit. The general physical and geological features of the auriferous belt of this state have been elaborately described by Clemson and R. C. Taylor,* and also by W. B. Rogers, State Geologist. The talcose slates, which predominate in the gold districts, have a reddish color, and are finely laminated, their general strike being from 29° to 32° east of north. The laminæ stand nearly vertically, and contain intercalated masses of syenitic granite and protogine. The breadth of the auriferous belt is about fifteen miles. For some depth these rocks are entirely decomposed, so as to be easily worked with the pick and shovel. The matrix of the gold is invariably quartz, which, near the surface, usually has a cellular structure, resulting from the decomposition and removal of the iron pyrites with which it was filled.

It was the opinion of W. B. Rogers, that many of these quartz veins might be worked with profit. In 1836, according to his authority, numerous quartz veins had been wrought for some time in Spottsylvania, Orange, Louisa, Fluvanna, and Buckingham Counties, from many of which rich returns had been procured, and, under improved modes of operation, a still larger profit might be expected.

At the present time, the following companies are known to be working on a somewhat enlarged scale; the results of their operations, however, are not usually made public; and, indeed, but few of them have yet been got fully under way. Several English companies have purchased some of the most promising gold mines in the State, and have recently commenced working them, and it is to be presumed that they will thoroughly test the question of their permanent productiveness.

Culpeper Mine, on the Rapidan River, seventeen miles from Fredericks-

---

* Trans. Geol. Soc. Pa. i. 298.

† Report of the Geol. Reconnoissance of the State of Virginia, 1836, p. 67.

burg. In 1850, working twelve stamp-heads and two Chilian mills, with twenty-four men. Weekly expenses estimated at $120. Produce in seven weeks, 3,400 dwts.

Freehold Gold Mining Company, Orange County. This is an English company, formed in 1853, after examination of the property by Mr. Henwood. One vein, of twenty feet in width, is said to have been opened, and traced for a mile and a half.

Liberty Mining Company. (100,000 shares at £1.) This is an English Company, which has purchased the Vaucluse and Gryme's Mines, for which £50,000 was paid. The average yield of the ore is estimated at $8 per ton. Six shafts have been sunk, and preparations made for working a large quantity of ore. According to the Directors' report, made Sept. 30th, 1853, the amount of gold produced during the year, the mill running eighty days, was 556 oz. 6 dwts., of a fineness of 943½ thousandths. In December, 1853, the stamps were crushing 50 tons per day.

The vein or lode worked consists of five parallel bands of quartz, all bearing gold. Hydrated oxide of iron is the principal associated mineral, and the enclosing rocks are talco-micaceous slates. Previous to 1852, the mine had been worked in two open cuts, to the depth of 60 feet, 75 feet wide, and 120 feet long. From a description of these mines, published in 1848, the following information is extracted.

The Vaucluse Mine is situated in Orange County, a mile south of the Rapidan River, about seventeen miles from Fredericksburg. It was discovered in 1832, and, for a number of years, worked as a deposit mine, before any veins were discovered. There are numerous veins now known which are contained in talcose and other slates. The gold is contained in quartz, and in the adjoining slates, which are much mixed with decomposed iron pyrites. Sometimes the auriferous belt widens out to 30 or 40 feet. The present establishment was commenced in 1844 by an English Company, and extensive works were erected. A Cornish engine, of 120 horse power, was connected with six Chilian mills, and six batteries of stamps, of three stamp-heads each. The finer ores are ground in the Chilian mills, and the harder quartz rock is stamped. The amalgamation is effected by the "amalgamating bowls." The quantity of mercury annually consumed is stated at from 250 to 300 lbs. The amount of gold produced, however, is not given, nor the quantity of ore ground and stamped. The gold obtained is very fine, being from ·985 to ·990.

Gardiner Gold Mining Company, Spottsylvania County. Situated at the junction of the Rappahannock and Rapidan Rivers; but little work has ever been done here. A steam-engine is now erecting, intended to drive two of Gardiner's gold-crushing machines, which are estimated to crush and amalgamate 100 tons of rock per day, the expense being $3, and the yield of gold $12 50 per ton.

Marshall Mine, Spottsylvania County, on the Rappahannock River, twelve miles from Fredericksburg. This mine is said to be paying well. It is stated that $300,000 has been obtained up to this time, twenty hands being employed, and the depth of the workings being 100 feet.

Whitehall Mine, Spottsylvania County. According to Mr. Henwood, this mine is worked in a deep-blue clay slate, the veins bearing about southeast and northwest. Their principal veinstone is a hard, white quartz, sometimes marked with ferruginous stains. Occasionally they enclose large masses of slate. The gold is for the most part scantily, but pretty uniformly, scattered through the vein, perhaps more plentifully in the iron-tinged parts than elsewhere; and where the quartz is somewhat drusy, it has a tendency to a crystalline structure.

Native tellurium has been found here, according to Mr. Henwood,* disseminated through the quartz, like the gold, but in smaller quantity. Galena also occurs with these metals; and the association of ore in this part of the Virginia gold region is said by the same authority to resemble that of the Morro de San Vincente, in the province of Minas Geraes, Brazil. A telluret of bismuth, containing selenium, has also been described by Mr. C. Fisher, Jr.,† as occurring at this mine.

Waller Gold Mining Company. This mine is situated in Goochland County, nine miles from Columbia, and was taken up by an English company in 1853, under the advice of Professor Ansted, who recently examined and reported on it. According to that gentleman's statements, it appears that the property is traversed by the great auriferous belt of Virginia, which is made up of various schists and alternations of shale and quartz rock, more or less compact and crystalline. Some of the quartzite is hyaline and micaceous, and almost all the rocks are colored by oxide of iron. Distinct indications of gold have been found disseminated through the shales and quartz in almost every part of the property in which the experiment of panning has been tried, and especially near the veins, which have been proved to the depth of from five to thirty feet. The results are stated as having been very variable, and are not given; but they are reported as sufficiently encouraging to justify the recommendation that the mine should be opened and proved by working it to a considerable depth.

The estimated yield of the ore is £4 per ton. In October, 1853, the Company was erecting buildings and machinery preparatory to working the mine on an extensive scale.

Garnett and Moseley Mines, Buckingham County. These mines were purchased by an English company, some time since, and are now among the most extensively worked in Virginia. Three steam-engines, of from thirty to sixty horse power, have been erected, by which seventy-two stamps and other machinery are driven. At the time the present Company came into possession, several shafts had been sunk, the deepest of which was 110 feet, at which point the vein is said to be fifteen feet wide, and to be worth $20 per ton. These mines were taken up on the recommendation of Professor Ansted, who is said to have reported very favorably on their prospects. According to Mr. A. Partz,‡ there are five or six veins on the property, two of which are worked. The principal one has a course of north 35° east, and dips 40° to the southeast; and the other runs in nearly the same direction, and dips so as to unite with the first. About forty miners are employed, and sixty surface-hands.

* Eng. Mining Journal, No. 956. † Sill. Am. Jour. (2) vii. 282.

‡ Mining Magazine, ii. 378.

GEORGIA.—In this state the yield of gold, which once reached half a million of dollars per annum, has fallen off gradually, until, in 1853, the amount deposited at the Mints was only $58,896.

Efforts have been recently made to revive some of the abandoned mines. Among those taken up, are Moore's Mine, near Dahlonega, by a company called *The Georgia Gold Company*, and the Lawhorn Mine.

TENNESSEE AND ALABAMA.—These states have each produced a few thousand dollars annually of gold, during the last twenty years; but there is no reason to suppose that their future yield will ever be of any considerable importance.

NEW MEXICO.—In regard to New Mexico, it seems almost impossible to obtain any definite information, although it is known that gold is produced here in some quantity. The amount deposited at the Mints is very small, and varying from a few hundred to several thousand dollars. Dr. Wislizenus* describes two localities near Santa Fe, the Old and New Placer. The "Old Placer" is twenty-seven miles from that city, the predominating rocks being white and yellow quartzose sandstone, quartz, hornblende rock, sienite, and diorite. At the "New Placer," nine miles from the town, both washing and mining in the rock are carried on. Two mines were worked at that time. The vein of one of these is described by Dr. Wislizenus as being contained in sienite and greenstone, and having a quartzose and ferruginous gangue. The same description applies to the mines of the "Old Placer." Together, these washings and mines are said, by the same authority, to have yielded at various times from $30,000 to $250,000 per annum.

CALIFORNIA.—We come now to speak of a country whose golden wealth surpasses anything yet known to have been discovered, and which, in its influence on the march of the world, is to be ranked among the great events of modern times. Up to a very recent period, the Pacific side of our continent had remained an uninhabited and almost unvisited region, of which we knew almost nothing, save what might

* Memoir of a Tour to Northern Mexico, 1846–7, published by Congress, p. 24.

www.ingramcontent.com/pod-product-compliance
Lightning Source LLC
LaVergne TN
LVHW011120110826
845150LV00008B/2197

* 9 7 8 1 4 2 5 5 5 7 0 6 5 *